COGNITIVE COMPUTING USING GREEN TECHNOLOGIES

COGNITIVE COMPUTING USING GREEN TECHNOLOGIES
Modeling Techniques and Applications

Edited by
Asis Kumar Tripathy, Chiranji Lal Chowdhary, Mahasweta Sarkar,
and Sanjaya Kumar Panda

CRC Press
Taylor & Francis Group
Boca Raton London New York

CRC Press is an imprint of the
Taylor & Francis Group, an **informa** business

First edition published 2021
by CRC Press
6000 Broken Sound Parkway NW, Suite 300, Boca Raton, FL 33487-2742

and by CRC Press
2 Park Square, Milton Park, Abingdon, Oxon, OX14 4RN

© 2021 Taylor & Francis Group, LLC

CRC Press is an imprint of Taylor & Francis Group, LLC

The right of Asis Kumar Tripathy, Chiranji Lal Chowdhary, Mahasweta Sarkar, and Sanjaya Kumar Panda to be identified as the authors of the editorial material, and of the authors for their individual chapters, has been asserted in accordance with sections 77 and 78 of the Copyright, Designs and Patents Act 1988.

Reasonable efforts have been made to publish reliable data and information, but the author and publisher cannot assume responsibility for the validity of all materials or the consequences of their use. The authors and publishers have attempted to trace the copyright holders of all material reproduced in this publication; we apologize to copyright holders if permission to publish in this form has not been obtained. If any copyright material has not been acknowledged, please write, and let us know so we may rectify it in any future reprint.

Except as permitted under US Copyright Law, no part of this book may be reprinted, reproduced, transmitted, or utilized in any form by any electronic, mechanical, or other means, now known or hereafter invented, including photocopying, microfilming, and recording, or in any information storage or retrieval system, without written permission from the publishers.

For permission to photocopy or use material electronically from this work, access www.copyright.com or contact the Copyright Clearance Center, Inc. (CCC), 222 Rosewood Drive, Danvers, MA 01923, 978-750-8400. For works that are not available on CCC please contact mpkbookspermissions@tandf.co.uk

Trademark notice: Product or corporate names may be trademarks or registered trademarks and are used only for identification and explanation without intent to infringe.

Library of Congress Cataloging-in-Publication Data

Names: Tripathy, Asis Kumar, editor. | Chowdhary, Chiranji Lal, 1975- editor. | Sarkar, Mahasweta, editor. | Panda, Sanjaya Kumar, editor.
Title: Cognitive computing using green technologies : modeling techniques and applications / edited by Asis Kumar Tripathy, Chiranji Lal Chowdhary, Mahasweta Sarkar, and Sanjaya Kumar Panda. Description: First edition. | Boca Raton, FL : CRC Press/Taylor & Francis Group, LLC, 2021. | Series: Green energy and technology : Concepts and applications | Includes bibliographical references and index.
Identifiers: LCCN 2020038792 (print) | LCCN 2020038793 (ebook) | ISBN 9780367487966 (hardback) | ISBN 9781003121619 (ebook)
Subjects: LCSH: Soft computing. | Machine learning. | Energy conservation--Data processing. | Electronic digital computers--Energy conservation.
Classification: LCC QA76.9.S63 C55 2021 (print) | LCC QA76.9.S63 (ebook) | DDC 006.3--dc23
LC record available at https://lccn.loc.gov/2020038792
LC ebook record available at https://lccn.loc.gov/2020038793

ISBN: 978-0-367-48796-6 (hbk)
ISBN: 978-0-367-63982-2 (pbk)
ISBN: 978-1-003-12161-9 (ebk)

Typeset in Times
by MPS Limited, Dehradun

Table of Contents

Preface vii
Contributors ix
About the Book xi
Green Engineering and Technology: Concepts and Applications xv

Part I Introduction

Chapter 1 Green Communication Technology, IOT, VR, AR in Smart Environment 3

Riyazveer Singh, Sahil Sharma, and Vijay Kumar

Chapter 2 Green Computing – Uses and Design 19

Rajendran Sindhu and P. Srividya

Part II Analysis

Chapter 3 Statistical Methods for Reproducible Data Analysis 37

Sambit Kumar Mishra, Mehul Pradhan, and Rani Aiswarya Pattnaik

Chapter 4 An Approach for Energy-Efficient Task Scheduling in Cloud Environment 59

Mohapatra Subasish, Hota Arunima, Mohanty Subhadarshini, and Dash Jijnasee

Chapter 5 Solar-Powered Cloud Data Center for Sustainable Green Computing 71

Saravanakumar A. and Sudha M. R.

Chapter 6 State-of-the-Art Energy Grid with Cognitive Behavior and Blockchain Techniques 95

R. Rajaguru, S. Praveen Kumar, and R. Krishna Prasanth

Chapter 7 Optimized Channel Selection Scheme Using Cognitive Radio Controller for Health Monitoring and Post-Disaster Management Applications 111

R. Rajaguru and K. Vimaladevi

Chapter 8 TB-PAD: A Novel Trust-Based Platooning Attack Detection in Cognitive Software-Defined Vehicular Network (CSDVN) 133

Rajendra Prasad Nayak, Srinivas Sethi, and Sourav Kumar Bhoi

Chapter 9 Analysis of Security Issues in IoT System 147

Likhet Kashori Sahu, Sudibyajyoti Jena, Sambit Kumar Mishra, and Sonali Mishra

Chapter 10 Resource Optimization of Cloud Services with Bi-layered Blockchain 171

J. Chandra Priya and Sathia Bhama Ponsy R. K.

Chapter 11 Trust-Based GPS Faking Attack Detection in Cognitive Software-Defined Vehicular Network (CSDVN) 191

Rajendra Prasad Nayak, Srinivas Sethi, and Sourav Kumar Bhoi

Part III *Applications*

Chapter 12 Cognitive Intelligence-Based Framework for Financial Forecasting 205

Sarat Chandra Nayak, Sanjib Kumar Nayak, Sanjaya Kumar Panda, and Ch. Sanjeev Kumar Dash

Chapter 13 Benefits of IoT in Monitoring and Regulation of Power Sector 223

Harpreet Kaur Channi

Chapter 14 Online Clinic Appointment System Using Support Vector Machine 239

Ch. Sanjeev Kumar Dash, Ajit Kumar Behera, and Sarat Chandra Nayak

Chapter 15 Electric Vehicle: Developments, Innovations, and Challenges 253

Sahil Mishra, Sanjaya Kumar Panda, and Bhabani Kumari Choudhury

Chapter 16 Usage of Convolutional Neural Networks in Real-Time Facial Emotion Detection 259

Ch. Sanjeev Kumar Dash, Ajit Kumar Behera, Sarat Chandra Nayak, and Satchidananda Dehuri

Index 275

Preface

Sometime in 2018, Asis and the rest of us, who have been collaborating on various research projects and publications over the years, wanted to congregate some of the cutting edge work that is being conducted in areas like virtual reality, augmented reality, blockchain, and internet-of-things. We realized that a major part of future research will encompass these areas of study — that was what led us to consider and uphold these particular topics — but we also realized that the current research work in these areas are as diverse as their applications. Therefore, we decided to focus on a fundamental concept that coherently threaded these apparently diverse research work — namely the computation and the energy efficiency aspect of these stellar research findings and ventures. Specifically, we wanted to bring together ideas, innovations, and lessons associated with cognitive computing and green computing and their application in diverse areas.

Thus evolved our book — *Cognitive Computing Using Green Technologies: Modeling Techniques and Applications.* Our aim has been to present a wide range of cognitive and green computing concepts and applications in the research verticals of IoT, AR, and VR environments. The 16 chapters that comprise this book dwell on topics and research findings on reproducible data analysis, solar powered cloud data center, security issues, resource optimization, cognitive computing in cancer imaging, and so on. These topics are likely to be embedded into cognitive computing environments for different applications. Cognitive computing performs an increasingly important role in modern data processing, information fusion, communication, network applications, real-time process control, and parallel computing. Additionally, green computing as a subset of cognitive computing will enable us to engage in eco-friendly use of the computing resources.

Our book has provided an interdisciplinary platform for researchers, educators, and practitioners to present and showcase their most recent innovations, research outcomes, challenges encountered, and solutions adopted in the field of green computing and cognitive computing. We hope that our readers will gain insight and deeper appreciation of where technology is headed in the future.

The Editors Team
July, 2020

Contributors

Dr. Asis Kumar Tripathy received his PhD in Computer Science and Engineering from the National Institute of Technology, Rourkela, India, and his MTech in Computer Science and Engineering from the International Institute of Information Technology, Bhubaneswar, India. Currently, he is working as an Associate Professor in the School of Information Technology and Engineering, VIT Vellore, India. He has authored more than ten national and international research papers to his credit. He is acting as a reviewer for some prestigious journals like Computers and Electrical Engineering (Elsevier), IET Networks, Wireless Personal Communications (Springer), Health and Technology (Springer), and International Journal of Biometrics (Inderscience). He is associated with many professional bodies like OITS, IACSIT, CSTA, and IAENG. His current research interests include Wireless Sensor Networks, IoT, and Computational Intelligence.

Dr. Chiranji Lal Chowdhary is an Associate Professor in the School of Information Technology & Engineering at VIT University, where he has been since 2010. He received a B.E. (CSE) from MBM Engineering College at Jodhpur in 2001, and M.Tech. (CSE) from the M.S. Ramaiah Institute of Technology at Bangalore in 2008. He received his Ph.D. in Information Technology and Engineering from the VIT University Vellore in 2017. From 2006 to 2010 he worked at M.S. Ramaiah Institute of Technology in Bangalore, eventually as a Lecturer. His research interests span both computer vision and image processing. Much of his work has been on images, mainly through the application of image processing, computer vision, pattern recognition, machine learning, biometric systems, deep learning, soft computing, and computational intelligence. As of 2020, Google Scholar reports over 300 citations to his work. He has given few invited talks on medical image processing. Professor Chowdhary is editor/co-editor of 3 books and is the author of over forty articles on computer science. He filed two patents deriving from his research.

Dr. Mahasweta Sarkar is an Associate Professor in the Department of Electrical and Computer Engineering at San Diego State University (SDSU), California, USA. Dr. Sarkar received her PhD in Computer Engineering from the University of California at San Diego (UCSD) in 2005. Her research interest lies in the area of wireless data networks. Her work addresses issues like scheduling, routing, optimal resource allocation, power management in wireless networks (WLANs, WMANs, Sensor Networks, Ad Hoc Networks), and wireless health. She has over 70 published research articles in technical journals, conference proceedings, and book chapters. She is the Director of the Wireless Networks Research Group at SDSU, where she leads a team of PhD and master's students, along with post-doctoral fellows, visiting faculty, and PhD students from India, China, Denmark, Iran, and Korea. She is the recipient of the President's Leadership award at

SDSU in 2010 for her excellence in research and the Outstanding Faculty Award in 2014 for her excellence in teaching.

Dr. Sanjaya Kumar Panda is working as an Assistant Professor and Head of the Department, CSE at IIITDM Kurnool, Andhra Pradesh, India. He worked as an Assistant Professor in the Department of IT at VSSUT, Burla, Odisha, India. He received his PhD from IIT (ISM) Dhanbad, Jharkhand, India; MTech from NIT, Rourkela, Odisha, India; and BTech from VSSUT, Burla, Odisha, India in CSE. He received two silver medal awards for the best graduate and the best postgraduate in CSE. He also received institutional awards like IEEE Brand Ambassador, SGSITS National Award for the Best Research Work by Young Teachers of Engineering College for the year 2017, Faculty with Maximum Publishing in CSI Publications Award, Young IT Professional Award (2017 and 2016), Young Scientist Award, CSI Paper Presenter Award at an international conference, and CSI Distinguished Speaker Award. He has published more than 60 papers in reputed journals and conferences. He is a member of IEEE, an associate member of IEI, Life Member of ISTE, and Life Member of CSI, IAENG, IACSIT, UACEE, ACEEE, and SDIWC. His current research interests include recommender systems, cloud computing, big data analytics, grid computing, fault tolerance, and load balancing.

About the Book

This book is organized into 16 chapters. Chapter 1 discusses the optimization of green communication, IOT, AR, and VR for a smarter environment. Telecommunication convention has touched novel statures in approaching eons and will be growing in the forthcoming years when it comes to wireless communication. The current technology of IOT has been freshly protracted to embrace niftier and additional operative handler communications. Various techniques such as invasion recognition system-based and anomaly-based invasion detection-based arrangements have been discussed in this chapter.

Chapter 2 reviews related literature to identify approaches related to green computing technologies. This chapter discusses factors affecting the environment, the need for green computing, uses of green computing, challenges faced, and the measures taken for green computing. Green Design focuses on planning energy-efficient computers, servers, printers, projectors, and other virtual gadgets that are environmental-friendly. Green Computing or Green IT is the study and practice of using computing resources in an eco-friendly manner to tone down the environmental impacts of computing and reduce unnecessary energy consumption using Artificial Intelligence, which has become a major topic of concern today.

Chapter 3 explores the data reproducibility through heuristic approaches directed towards data handling. The chapter is also aimed towards easing out the modeling of sets of data. As a profession and research, it still has many areas that could unleash useful methodologies that may cover and help in projects and result in new algorithms.

Chapter 4 presents a dynamic green optimized energy scheduling using a dynamic load balancing technique. Here, VMs are selected according to the SLA level given by the user (i.e. deadline).

Chapter 5 deals with recent research focuses on the challenging issues of energy consumption in mobile cloud computing. As technology advances, the need for energy increases, including all physical and non-physical entities in mobile devices and advanced computers, such as servers and desktops.

Chapter 6 explores energy uncertainties which can be analyzed with time series analysis techniques in machine learning to estimate the peak hours for both prosumers and consumers. Prosumers can effectively participate in the grid with smart contracts. Combining multiple grids can even reduce maintenance outages.

Chapter 7 describes the Cognitive Radio Network (CRN) as a promising solution to prevent spectrum scarcity and improve spectrum utilization dramatically. The CRN stays as an optimal solution to cut down improper communication without delay, as proposed in this presented work. The information will be collected and transferred to the central node called cognitive controller from disaster or rural regions.

Chapter 8 discusses a trust-based platooning attack detection (TB-PAD) method proposed for cognitive software defined vehicular network (CSDVN). In this method, Road Side Units (RSUs) at the junctions receive the information to detect the platoon. This method mainly uses cognitive computing technology through prior knowledge of the vehicles to generate new trust values.

Chapter 9 focuses on IoT security which deals with a large amount of confidential data. It is essential to make IoT secure to guarantee client satisfaction. Applying security countermeasures on an IoT framework is difficult due to the diverse type of devices used and its pervasive nature. As most cryptographic algorithms are not compatible with IoT devices due to low memory and computing resources, new lightweight cryptography is used to enable a varied range of modern security applications for heterogeneous devices.

Chapter 10 reviews blockchain as an immutable distributed ledger that can be used to optimize the storage as objects are redundantly stored on multiple devices across multiple facilities. It can confirm a person accessed a specific piece of information by logging the hash of it that is hooked on a blockchain. It also urges the protection of top-secret and mission-critical data through decentralized management and access in cloud-based systems.

Chapter 11 exemplifies a trust-based GPS faking attack detection method proposed for cognitive software defined vehicular network (CSDVN). In this method, the vehicles in the network use the received signal strength indicator (RSSI) value of the beacon signals of neighbor vehicles to detect malicious activity.

Chapter 12 explains the most favorable ANN structure deployed for modeling and forecasting ten financial time series datasets in real-time, including indexes of daily stock index closing, foreign exchange training, and crude oil price. The suitability of evolutionary optimization-based artificial neural network models is discovered by systematic analysis and comparative study.

Chapter 13 exemplifies the main role of IoT in the electrical sector. IoT helps people and the government, through Smart Grid technologies and smart city solutions, to improve and replace older architecture.

Chapter 14 focuses on reducing the waiting time in clinics. They provide a scheduling procedure which is a combination of machine learning and mathematical programming. Determining the priority of outpatients and allocating the capacity based on priority classes are important concepts that must be considered in scheduling outpatients. They applied clustering methods such as k-mean clustering to classify outpatients into priority classes. and suggested the best pattern to group them.

Chapter 15 introduces electric vehicles (EV) with the challenges faced to develop them. Now, electric vehicles have become a priority for governments and automobile manufacturers. A lot of hurdles are in the development of EVs — frequent power cuts in various regions of the world — which results in deterioration of the performance of the vehicles.

Chapter 16 explores a novel CNN technique for facial emotion recognition. A video stream input stream is inputted to investigate the above-said problem using CNN. İn this context, they have developed a framework to detect human emotions, different directions, and lighting conditions in real-time.

We would like to convey our earnest appreciation to all the authors for their contributions to this book. Starting from the call until the finalization of chapters, all contributing authors have given their input amicably, which is a positive sign of significant teamwork. We would like to extend our gratitude to all the reviewers for their constructive comments on all the chapters. We are very much thankful to series editors Brojo Kishore Mishra and Raghvendra Kumar for their input for this book. Finally, we would like to thank all the members of the CRC Press/Taylor & Francis Group for providing constructive inputs and allowing an opportunity to edit this important book.

Green Engineering and Technology: Concepts and Applications

Series Editors: Brujo Kishore Mishra, GIET University, India and Raghvendra Kumar, LNCT College, India

Environment is a critical issue these days for the world. Different strategies and technologies are used to save the environment. Technology is the application of knowledge to practical requirements. Green technologies encompass various aspects of technology which help us reduce human impact on the environment and creates ways of sustainable development. Social equability: this book will enlighten green technology in different ways, aspects, and methods. This technology helps people understand the use of different resources to fulfill needs and demands. The combination of involuntary approaches, government incentives, and a comprehensive regulatory framework will encourage the diffusion of green technology in developing countries.

Green Innovation, Sustainable Development, and Circular Economy

Edited by Nitin Kumar Singh, Siddhartha Pandey, Himanshu Sharma, and Sunkulp Goel

Green Automation for Sustainable Environment

Edited by Sherin Zafar, Mohd Abdul Ahad, M. Afshar Alam, Kashish Ara Shakeel

AI in Manufacturing and Green Technology

Methods and Applications

Edited by Sambit Kumar Mishra, Zdzislaw Polkowski, Samarjeet Borah, and Ritesh Dash

Green Information and Communication Systems for a Sustainable Future

Edited by Rajshree Srivastava, Sandeep Kautish, and Rajeev Tiwari

Handbook of Research for Green Engineering in Smart Cities

Edited by Kanta Prasad Sharma, Abdel-Rahman Alzoubaidi, Shashank Awasthi, Ved Prakash Mishra

Green Internet of Things for Smart Cities

Concepts, Implications, and Challenges

Edited by Surjeet Dalal, Vivek Jaglan, and Dac-Nhuong Le

Green Materials and Advanced Manufacturing Technology

Concepts and Applications

Edited by C. Samson Jerold Samuel, M. Suresh, Arunseeralan Balakrishnan, and S. Gnansekaran

Cognitive Computing Using Green Technologies

Modeling Techniques and Applications

Edited by Asis Kumar Tripathy, Chiranji Lal Chowdhary, Mahasweta Sarkar, and Sanjaya Kumar Panda

Multiple Objective Analytics for Criminal Justice Systems
Gerald W. Evans

For more information about this series, please visit: https://www.routledge.com/Green-Engineering-and-Technology-Concepts-and-Applications/book-series/CRCGETCA

Part I

Introduction

1 Green Communication Technology, IOT, VR, AR in Smart Environment

Riyazveer Singh[1], Sahil Sharma[1], and Vijay Kumar[2,*]

[1]Thapar Institute of Engineering and Technology, Patiala, Punjab, India
[2]CSED, National Institute of Technology, Hamirpur, Himachal Pradesh, India

1 INTRODUCTION

Numerous objects like smart conveyance, nifty metropolitan and self-directed automobiles, etc. have come by with the growth of IoT tenders. Ultra-low latency communication for mobile phones and other gadgets (e.g., smartwatches) are essential to achieve system reliability—real-time analysis of traffic by smart vehicles to avoid accidents, smart parking systems, smart lighting systems, smart dustbins with automatic trash segregation, etc. Enabled by machine learning (ML) and self-governing maneuvers via superiority campaigns or nifty apparatuses/representatives, upbeat web connotation is perilous in circulated calculating circumstances—autonomous automobiles driving and backup liberation—by eradicating dependence on the central executive unit. AR and IoT have conventional significant consideration of significant skills [1,2] meant to assemble imminent active sitting room, further reception, and extra communication— everchanging our existence. AR remains a form of collaborating standard that delivers an interpretation of the physical creation amplified by—in addition/or spatially itemized—valuable cybernetic data. This authorizes manipulators to comprehend the creation and intensify their intellect in resolving glitches and guiding numerous errands [3]. Innovative AR bids a favorable method for handlers to envisage and intermingle through corporal things and their associated information. Multifaceted community structures comparable to airfields use various systems to lead the public to a firm endpoint. Such methods are characteristically executed by bestowing a level strategy, partaking organizational symbols, or color oblique outlines on the sordid. It is feasible to find out the trusty persons by using augmented reality mechanism, which have provision of anomaly tracing in the interior surroundings. All the facets of networks—peripherals, routers, switches, communication media, and personal computers—are sheltered by green networking.

The global energy consumption will be pointedly influenced by the optimization of liveliness efficacies of all web workings. Subsequently, these expanded by having a Jade Grid [4–7] will diminish CO_2 productions and help confine international heating. Internet of Things (IoT) is a fresh movement beneficial to diverse submissions spheres. Smart tenders such as smart buildings, smart well-being, a keen environment, and smart metropolises can be shaped by an IoT infrastructure. Integrating computer hardware and software system appliances in smart atmospheres can be advantageous, though the cost of using IoT is indispensable when it comes to choosing IoT as a technology to enable a smart environment. Consequently, cost operative ways of using IoT is decisive from the standpoint of both corporate owners and clients.

This tabloid describes numerous ideas connected to IoT and MIoT, such as the Mobile Internet of Things (MIoT), Smart IoT Environment, MIoT Shareholder, MIoT Scenario Application, and MIoT Facility. The descriptions of this terminology are "trails". Mobile Internet of Things (MIoT) is an IoT scheme that can be stimulated and reprocessed in diverse places. A smart IoT situation is a town or structure built, rebuilt, or grounded on the IoT system; MIoT stakeholders are staff working on the construction and maintenance of the MIoT system. MIoT Landscape Request is an ideal claim that can profit from the MIoT system. The MIoT facility is a software program facility that assists the jobs of the MIoT system (Figure 1.1).

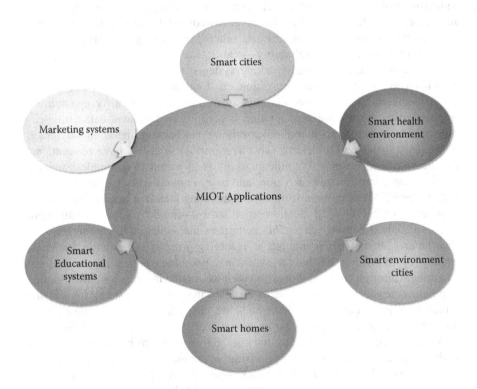

FIGURE 1.1 Propositions of MIoT.

Green Communication Technology

The remaining sections of the proposal cover related work in Section 2 and techniques in Section 3. Section 4 presents smart city applications. The conclusion is presented in Section 5.

2 ASSOCIATED WORK

Evaluation of connected research highlights three main mechanisms and necessities of our planned IoT + AR-style—the current state on AR/IoT data organization, preceding strategy for interacting with IoT objects, some use cases with AR, and normal content illustration or organization interoperability etiquettes. AR Data and Content Description for Corporal Goods AR Services typically maintain corporeal [8] daily data and service content for their components or improvement goals (Figure 1.2).

Therefore, we consider current methods for such corporeal object data (for example, construction and information management) for AR use. The previous AR system was used as a solo claim with all its content and assets, using programming libraries. For example, the augmented content of an object is easy (only shows growth potential) and inorganic. Mobile and smartphones with GPS allow the establishment of position-based geographic and AR facilities to provide insights to tourism and strategies for viable points. Such a feature is engaged to support content (and its setup stipulations); the perception of content and provision everywhere, as well as the united monitoring of data on the mainframe-server HTML, KML, [9] and ARML markup dialects for such persistence. For example, Vicki predicts Augmented Reality Mark-up Language [10] (ARML) for position-based facilities. ARML permits you to define geographic ideas or sites of attention and add GPS coordinates and simple improvement materials (for example, transcript, symbols, and imageries). There are many other contented depiction approaches for AR facilities, but they need complex scripting for an explicit submission or contented category (e.g., cinematic built, AR online manuals, and AR guides), or are deprived of adequate capture. Though, a typical interdisciplinary content format is

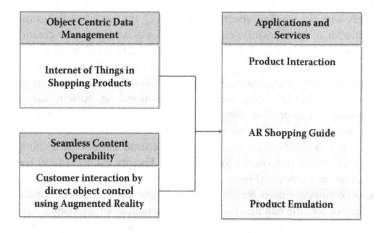

FIGURE 1.2 Analysis of Smart Shopping Experience.

proposed to represent different types of AR services. Also, billions of corporeal stuffs can mechanically connect with processers and an alternative for mass smart facilities. In this case, it is important to name ascendable substances and specify typical data arrangements with AR-based amenities—innovation of Smart City Information Technology and Information Economy. It is a continuation of urban development, scientific generation C observation, novel learning of information technology, full understanding of data and the internet, self-sensing, intelligence, self-adaptation, and visual collaboration—and to extend public life, eco-friendly fortifications, religious security, urban businesses, corporate activities, and other municipal demands for diversity of smart retailers, harmless opportunity and internal mobility, and sustainable development of EF-Cent Green City. Implementing a smart city is the adaptation of a public system, the latter being more vulnerable, smarter, more harmonious, and more sophisticated, while the public and goods of the city are extra smart and melodious and can animate opportunely subsequently. As VR (virtual reality) specialization becomes more widespread and emergent in adults, the mandate for VR [11] also increases. Aside from Virtual Reality Technology and Terrestrial Statistics Structure, VRGIS (Virtual Reality Geographic Information System) is also an advanced data organization. Through the commercialization of the Internet, grid-based VRGIS [12] has become increasingly popular. In this case, VR system networking is attained via the submission of online VR knowledge [13] like X3D and VRML. However, operational VR expertise [14,15] motionlessly faces many challenges for large-scale information, system bandwidth transmission restraints, substantial needs, and multi-user collaborative controller. To upsurge the accuracy of modeling, VR systems in urban planning [16] require more realistic performance. Most importantly, this situation can cause data transmission to increase.

Additionally, we are studying the ideas that we endorse to inspect at solitary cell and node stages.

Due to the epoch of green message, we may bargain in the field of intellectual wireless communication. The device understands its atmosphere, adapts to diverse circumstances, and has long-term effort and establishment. Integrating the needs of system users with the wireless spectrum range [17], the use of power grids and the power controller of base position equipment unlocks an innovative creation of intellectual governor. A cerebral process can be linked to an expansion path, and a visual turf can be used to visualize the loss of a path in a given location due to altering climate situations (shower, mist, or vegetation vicissitudes). It is necessary to emphasize the cross-layer approach to non-traditional, interdisciplinary strategy, where improvement to prevent component-level interactions is seen but leads to a net-zero gain or a net loss. The Power Optimization Context, based on the design of Cerebral Radio, is established and integrates scheme metrics to enthusiastically enhance system usage in terms of Superiority of Facility (SOF) [18,19] for precise radio situations and tenders (Figure 1.3).

The most commonly used way is to interpret and connect cardinal matters. The custom edges denote the handheld isolated panels such as smartphones. AR delivers rigorous improvements over object recognition and documentation process. Though a server-based method to omnipresent AR facilities is possible with ordinary

Green Communication Technology

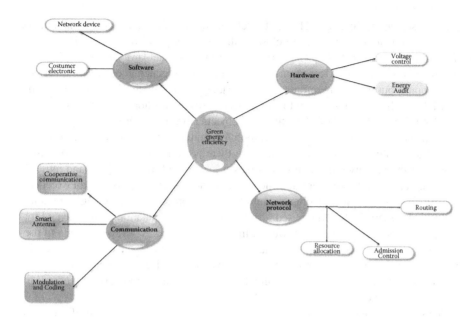

FIGURE 1.3 Fields with Energy Competence System Necessities for Enhanced Sustainability and Abridged Rate.

corporeal substances, it is difficult to obtain them for big average places due to object recognition procedure, complex feature corresponding, and contented lookup for multiple things. There are high recital cloud computing amenities for debauched object matching processes and accelerate content recovery related to AR facility delivery. However, it is problematic to raise the scalability "everywhere," side by side. The substitute is to link a limited zone server (only a specific resident zone, like a solitary home-based service) to handle a restricted quantity. This is comparable to the idea of fog calculating to permit calculating facilities of head-to-head networks for efficient information organization.

Rathore and Park [20] offered a fog-based bout recognition agenda to perceive bouts in IoT. This method has been recommended to address unworthy outcomes in federal bout discovery systems due to scalability, supply, reserve constraints, and truncated dormancy. Sharma et al. proposed fog bulge construction to reduce safekeeping bouts for physical period analytics facilities. Therefore, a sifting mechanism (to decrease hunt area) is proposed, such as transmitting communications for near users. Iglesias et al. recommended a technique to classify and improve applicant board substances with relevant information such as operator quality, operator article immediacy, comparative alignment, reserve discernibility, and concluding assortment physically. Others involved in Azamiki projected a parallel idea. Regrettably, there is not at all significant effort involved in clambering AR facilities and their well-organized information organization in a mostly ordinary environment. AR services can be utilized at home, office, street, or shopping area. The above procedures are grounded on the essential system server style, as already specified, and can impede a stern presentation blockage [21,22]. Therefore, some studies attempt to obtain datasets directly from objects that are close to the user. As previously stated in our framework, AR-enabled IoT devices now store (internal) standard "features" that can be used to track mobile AR customers and communicate general content for a variety of shopping purposes. Therefore, communicating control of the IoT device is conceivable. With a filtering system to choose from between millions of different objects, the AR Client will similarly find that IoT objects (fortified with basic dispensation, storing, and system components) detect mobile admittance points, providing the client with the essential AR pursuing data. Since there are only a small number of target objects in the vicinity, AR clients can quickly locate (and track) objects and retrieve relevant content. AR's current and most prevalent application provides an excellent control method for in situ object control (or for remote objects using a remote-controlled camera). AR can be used to visualize control simulations implemented for preview or training purposes. However, there have been only a few attempts to use AR (or VR) as a control and simulation interface. As a result of the previously intended results, Rekimoto and Ayatsuka have proposed a visual tagging system called cyber codes, which uses 2D barcodes to detect objects, and various methods to manipulate physical objects (Figure 1.4).

This was verified in a congested space at the Vienna airport, about 200 meters long (around 2,000 square meters). When switching between individual maps, we have achieved continuous tracking with a standard tablet in just one millisecond (100 m–300 m). The FOVPath concept has been executed as a plugin for Unity 3D.

More recently, Als et al. projected an assessment of three elementary models of AR interactions to control Io environments [23,24] (floating icons, floating menus, and WIM) and found that the WIM model is complex and time-consuming. In our case, we have explored AR's affiliate mechanism by a smartphone with preconfigured information that connects to IoT products [25]. Although it is possible for such AR connections (for example, on common remote-controlled species), it is unclear how to establish coherent AR-mediated interactions with

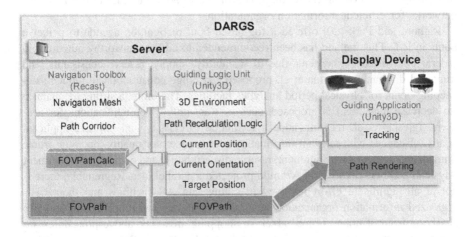

FIGURE 1.4 Working of Dynamic AR Guiding System.

lots of dissimilar elements. In the earlier trainings cited above, AR compensation was largely constructed in a temporal manner.

Use circumstance scenario: spending for IoT + AR [26,27], we show two scenarios using AR-enabled hardware, highlighting the functionality of our IoT-enabled AR system [28]—the three main components described above. Highlights: object-to-object data management, object access, control, and object interaction, as well as interactive content exchange.

Samila enjoys supermarket run. Wearing his AR glasses, he goes to a closed electronic store to purchase a speaker. Since there are so many speakers on display, you are too embarrassed to choose. He links with these speakers, and his glasses reveal a wide variety of product-related information (e.g., price, variety, flexibility, and availability) integrated directly into the products. He still hesitates and decides to keep up with his audio class. She writes a specific prototype using her finger (followed by a camera with AR Glass) and then calls Samila to insert her own MP3 file to test the model. The mirrors show the speaker's position and the soundwave propagation simulates the effects of sound and the resulting musical quality. In addition to listening, gathering, and displaying useful information, IoT objects [29,30] are "statistically controlled" to find relevant resources. In many cases, the direct RR target is required, and the AR is an appropriate indicator (for example, in comparison to the switch key crossing point) because it delivers the data needed to make the chore calmer and more convenient [31]. For example, IoT devices [13] can be integrated into the system sensor [32,33] by adding computing power. Therefore, objects can transmit vital information based on the customer's visual information (including information and data needed for the following). That is, information is now transmitted to separate objects in the setting. Consequently, our method delivers the best and most expected substructure to access "universal" AR to virtual objects. Note that information and/or content can be uploaded to add-ons and novel shopping apps are created. In addition to general data and in-app content, AR clients can communicate the data needed to view, visualize, and engage with each item of interest, including features and purchase conditions. The security of IoT systems is a major concern due to the increasing number of services and customers in the network. The combination of IoT systems [34,35] and smart environments makes smart objects more efficient. However, the effects of exposure to IoT security in smart environments used in fields such as medicine and manufacturing are particularly dangerous. A smart IoT-enabled environment [36] that does not have full security, systems, usage, and services are at risk. Privacy, authenticity, and readiness are the three most important security elements for upland services in an IoT-based smart environment. Therefore, further research on information security in IoT systems is needed to address these issues. For example, IoT-based smart homes [37] face security and privacy challenges that cover all layers of the IoT style. Creating smart surroundings in the real world faces two significant obstacles: the security of IoT systems and the complexity and compatibility of IoT environments [38,39]. DoS or DDoS attacks on IoT networks [40] affect IoT services and, consequently, services provided through smart environments. Investigators have studied the security challenges of IoT from many different aspects, one of which is the security susceptibility of the IoT announcement protocol (Figure 1.5) [41].

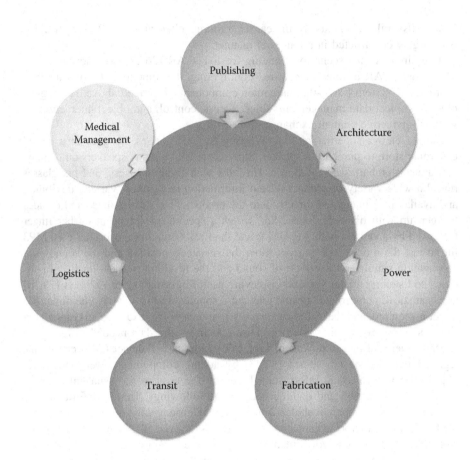

FIGURE 1.5 Scope of IoT-Based Smart Environment.

Safety tests in IoT schemes [42] are connected to safety matters from dissimilar areas of IoT [43]. Corporeal damage, computer hardware failures, and authority restrictions are encounters before carnal coating. DoS bouts [44], hacking, gate outbreaks, and illegal admittance are encounters related to the system's scope. Malevolent cipher bouts, software susceptibilities, and software program pests are challenges that the submission layer faces. Accordingly, the safety connected issues of any IoT system [45,46] can be alienated into four types: verification and carnal intimidations, security dangers, information veracity subjects, and confidentiality issues.

3 TECHNICAL TECHNOLOGY

3.1 Strategies Based on Program Acquisition

The different stages of logging rely on detection system (IDS) algorithms. There is a large number of procedures for all IDS methods. Some of these are briefly discussed in the section entitled 'IoT Systems Intended for IDS' [47]. In addition, they can be used for various methods of discovery. Therefore, this section emphasizes wireless IDS algorithms that can be deployed in an IoT-based environment depending on intricacy, implementation time, and acquisition time necessities. Principal Component Analysis (PCA) [48] is a frivolous procedure that is used for numerous recognition methods in IDS and is deliberated as a practical example. Jolliffe and Cadima [49] stated that "Principal Component Analysis (PCA) is a widely cast-off multivariate evocative technique for managing numerical information and can be extended to deal with varied dimension level information." As labeled, PCA produces a set of variables contingent on the variance-covariance construction of unique variables. These new variables are linear groupings of the original variables and are less in number than the novel variables.

In IDS, PCA is used as a dimensionality decrease and discovery method. Elrawyetal used the PCA method to create an anomaly-based arithmetical and statistics withdrawal IDS that is contingent on the dissection of the principal mechanisms into the maximum and slightest important principal components. In this system, the recognition phase depends on the chief principal component score and the inconsequential principal component score. In accumulation, PCA has been used in intrusion detection techniques grounded on cargo modeling [50], arithmetical modeling, data mining, and apparatus erudition.

3.2 IRREGULARITY GROUNDED INTERRUPTION RECOGNITION

In the irregularity grounded interruption recognition method, a usual information design is shaped by information from standard operators associated to present statistics designs to observe irregularities that rise due to clamor (or other singularities that have roughly the likelihood of being formed by riding apparatuses). Therefore, irregularities are uncommon behaviors caused by interlopers that leave paths in the computing atmosphere. These paths are perceived to classify mainly unidentified outbreaks [51]. An irregularity grounded IDS makes a prototype of the standard behavior in the computing setting, which is uninterruptedly rationalized, grounded on information from usual operators. Using this prototype, any eccentricity from usual behavior is perceived.

3.3 IDSs: PRESENTATION ASSESSMENT

The actions used to evaluate IDS presentation hinges on four features: the numbers of true positives (α), true negatives (δ), false positives (γ), and false negatives (β). Following these factors, the performance metrics for IDS are described below. In predicting the anomaly class, a true positive (αA) is a correct classification that indicates an intrusion. A true negative (δA) is a correct classification that indicates no intrusion. A false positive (γA) is an incorrect classification that indicates an intrusion when there is none. A false negative (βA) is an incorrect classification that indicates no intrusion when there is one. The factual optimistic rate (FOR), which describes the probability of detecting intrusions is calculated as:

$$FOR = \alpha A/\alpha A + \beta A \qquad (1.1)$$

TABLE 1.1
Main Factors Considered in Intrusion Detection System (IDS)

Forecasting the Irregularity Period		Outbreak	Standard
Real	Outbreak	αA	βA
Projected	Standard	γA	δA
Expecting the Standard Period		Standard	Outbreak
Real	Standard	αN	βN
Predicted	Attack	γN	δN

The incorrect optimistic rate (IOR), which labels the probability of erroneously classifying normal behavior as an interruption, is intended as:

$$IOR = \gamma A/(\gamma A + \delta A) \tag{1.2}$$

The memory (R), which defines the proportion of the total pertinent archives in a catalog that are recovered by penetrating, is intended the way TPR is. The accuracy (A), which describes the proportion of pertinent archives among there cords recovered, is intended as (Table 1.1):

$$A = \alpha A/(\alpha A + \gamma A) \tag{1.3}$$

The F-score (F), which defines the equilibrium between A and R, is intended as:

$$F = 2_*A_*R/(A + R) \tag{1.4}$$

The general success rate, which defines the proportion of accurate organizations, is intended as:

$$\text{Success Rate} = \alpha A + \delta A/(\alpha A + \delta A + \gamma A + \beta A) \tag{1.5}$$

$$\text{Error Rate} = 1 - \text{Success Rate} \tag{1.6}$$

When envisaging the standard class, similar descriptions and calculations can be used, except with the constraints αN, βN, γN, and δN.

4 KEEN METROPOLITAN TENDERS

This section proposes instances of keen metropolitan tenders [52] for nifty transportation, useful energy, smart well-being, ambient abetted existing, misconduct deterrence and communal security, governance, disorder nursing and conservation

Green Communication Technology

of substructure, tragedy supervision and backup, canny households [53,54], travel and regeneration, and ecological organization.

4.1 Smart Transportation

A combined transport scheme would require a permit in the form of a smart card, which can be encumbered with cash and is swiped at any point of admission into a conveyance organization using Near Arena Message (NAM) expertise— communicating data from the card to the interpreting appliance and back. Reimbursement is subtracted from the card for the journeys completed. A separate parking cove is a meter that senses a car parked through a label on the number plates as soon as the car arrives at the bay and starts scheming the charges for the parking. Drivers catalog e-toll accounts with road agencies and are delivered with radio-frequency identifier (RFID)-enabled e-toll cards [55] attached to the cars. As the car determinations under an e-toll gate, the driver's particulars and the details of the distance they have traveled are read by the card reader on the e-toll entrance and transmitted to a server at the road's agency.

4.2 Ambient Abetted Existing

To sojourn aging for an elongated period in treatment homes, body sensors are used to detect body parameters. These sensors connect to their caregivers wirelessly. Should any of the parameters go out of range, an alarm to the caregiver is activated.

4.3 Corruption Deterrence and Public Security Documentation of Convicts

It has been made relaxed by itinerant biometric recognition machineries. Suspicious thumbprints are seized to a police itinerant biometric apparatus. This information is directed via a system to a fingerprint catalog situated at the Department of Home Affairs for assessment.

4.4 Ascendency

The numerous services accessible online, their functionality and the side-by-side use, and their level of communication are significant pointers of e-government's 'smartness' standards. Taxes, compost, energy, and tax revenues respectively take on a computer-labeled identity document and the computer is properly integrated to process the information on the record, and the costs and expenses are met.

4.5 Disruption of Nursing and Maintenance of Replacement

The devices recognize the operating technology of a traffic congestion account with a system that secures the physical integrity of the links after the bus passes.

4.6 Disaster and Backup Organization

The satellites have found the signatures of recently launched wildfire and transmit information to the fire control center and fire buses. A similar regulator center creates fire alarms that are in designated areas to alert residents.

4.7 Ecological Monitoring

Urban engineers are installing devices throughout the town that ration temperature, comparative moisture, carbon oxide, nitrogen monoxide, clatter, and elements. If one of the strictures is overhead the specified limit, GPS-empowered devices direct the concern to a non-core location. The node also directs data to residents' phones.

4.8 Waste and Equipment Administration

The city has instruments mounted in infected cisterns to increase terror when the infected cistern reaches set levels. Automobiles are sent to eradicate surplus. The city is putting drums in the right places in the metropolitan. Containers with devices increase apprehensions once the bin is full and a garbage automobile is sent to gather left-over.

4.9 Canny Is Taken Home

The households we live in can be adjusted to classify specific people—whether they are ordinary domestic associates, visitors, or unsanctioned people. People are recognized by what they have—for example, a smartphone that is exposed to wireless rollers. If an in-household person does not have an ID then they are recognized as unauthorized and the dreaded sign is directed to an outside agency or homeowners.

4.10 Smart Energy

A transportable app allows people to regulate their electronic devices. Operators choose an app from the Appstore and turn it off. Appeal to turn off GSM grid over to an IP app for home-based application.

5 CONCLUSION

The work presented looks at the rural destinations of Internet of Things (IoT) organizations in software and mimics parts of the created space. We have defined which existing AR organization can be secured to match multiple user interfaces in actual store environments. The presented work discusses related work in detail and highlights the examples and challenges in the IoT and the five urban claims. The discussion of smart city use emphasizes the scope of work, solving the challenges that are needed in a smart city. The role of virtual reality is discussed in how it can address the challenges of a good city.

REFERENCES

1. Gimenez, R. and Pous, M. (2010). Augmented reality as an enabling factor for the Internet of Things. In: W3C Workshop: augmented reality on the web.
2. Zhang, H. (2009). Green Communications and Green Spectrum, CHINACOM, Hangzhou, China.
3. Wagner, D. and Schmalstieg, D. (2003, October). Artoolkit on the pocketpc platform. In 2003 IEEE International Augmented Reality Toolkit Workshop (pp. 14–15). IEEE.
4. Patil, S., Patil, V. and Bhat, P. (2012). A review on 5G technology. *International Journal of Engineering and Innovative Technology (IJEIT)*, *1*(1), pp. 26–30.
5. MacCartney, G. R., Zhang, J., Nie, S., and Rappaport, T. S. (2013, December). Path loss models for 5G millimeter wave propagation channels in urban microcells. *In Globecom* (pp. 3948–3953).
6. Webb, M. (2008). SMART 2020: enabling the low carbon economy in the information age, a report by The Climate Group on behalf of the Global eSustainability Initiative (GeSI). *Creative Commons*.
7. Ko, H., Lee, J., and Pack, S. (2018). Spatial and temporal computation offloading decision algorithm in edge cloud-enabled heterogeneous networks. *IEEE Access*, *6*, pp. 18920–18932.
8. Tripathy, A. K., Das, T. K., and Chowdhary, C. L. (2020). Monitoring quality of tap water in cities using IoT. In *Emerging Technologies for Agriculture and Environment* (pp. 107–113). Springer, Singapore.
9. Nolan, D. and Lang, D. T. (2014). Keyhole markup language. In *XML and Web Technologies for Data Sciences with R* (pp. 581–618). Springer, New York, NY.
10. Lechner, M., (2010). Arml-augmented reality markup language. *Ginzkeyplatz*, *11*, p. 5020.
11. Chen, Y., Zhang, S., Xu, S., and Li, G. Y. (2011). Fundamental trade-offs on green wireless networks. *IEEE Communications Magazine*, *49*(6), pp. 30–37.
12. Cavagna, R., Bouville, C., Royan, J. (2006, November). P2p network for very large virtual environment. In Proceedings of the ACM Symposium on Virtual Reality Software and Technology (pp. 269–276).
13. Hu, J., Wan, W., Wang, R., and Yu, X. (2012). Virtual reality platform for smart city based on sensor network and osg engine. In IEEE 2012 International Conference on Audio, Language and Image Processing (ICALIP), pp. 1167–1171.
14. Huang, B., Jiang, B., and Li, H. (2001). An integration of GIS, virtual reality and the internet for visualization, analysis and exploration of spatial data. *International Journal of Geographical Information Science*, 15(5), pp. 439–456.
15. D. Koller, D., Lindstrom, P., Ribarsky, W., Hodges, L. F., Faust, N., and Turner, G. (1995, October). Virtual GIS: A real-time 3D geographic information system. In Proceedings Visualization'95 (pp. 94–100). IEEE.
16. Edwards, M. (2013, February 4). Virtual reality system including smart objects. US Patent App. 13/758,879.
17. Chen, Tao, Yang, Y., Zhang, H., Haesik, K., and Horneman, K. (2011, October). Network energy saving technologies for green wireless access networks‖ IEEE Wireless Communications.
18. He, J., Loskot, P., O'Farrell, T., Friderikos, V., Armour, S., and Thompson, J. (2010, August). Energy efficient architectures and techniques for Green Radio access networks. In *2010 5th International ICST Conference on Communications and Networking in China* (pp. 1–6). IEEE.
19. Bianzino, A. P., Chaudet, C., Rossi, D., and Rougier, J.-L. (2012). A Survey of Green Networking Research. *IEEE Communications Surveys and Tutorials*, *14*(1), first quarter.

20. Rathore, S. and Park, J. H. (2018). Semi-supervised learning based distributed attack detection framework for IoT. *Applied Soft Computing*, 72, pp. 79–89.
21. Weber, M. and Boban, M. (2016, May). Security challenges of the internet of things. In 2016 39th International Convention on Information and Communication Technology, Electronics and Microelectronics (MIPRO) (pp. 638–643). IEEE.
22. Čolaković, A. and Hadžialić, M. (2018). Internet of Things (IoT): A review of enabling technologies, challenges, and open research issues. *Computer Networks*, 144, pp. 17–39.
23. Gerstweiler, G., Vonach, E., and Kaufmann, H. (2015). HyMoTrack: A mobile AR navigation system for complex indoor environments. *Sensors (Switzerland)*, 16(1).
24. Gerstweiler, G., Kaufmann, H., Kosyreva, O., Schonauer, C., and Vonach, E. (2013). Parallel tracking and mapping in Hofburg Festsaal. In Virtual Reality (VR) (pp. 1–2). IEEE.
25. Want, R., Schilit, B., and Jenson, S. (2015, January). Enabling the Internet of Things, *Computer*, 48(1), pp. 28–35.
26. SPU (2005). The internet of things executive summary. Technical report. The ITU Strategy & Policy Unit, (SPU).
27. Ray, P. P. (2018). A survey on Internet of Things architectures. *Journal of King Saud University Computer and Information*, 30(3), pp. 291–319.
28. IEEE Minerva, R., Biru, A., and Rotondi, D. (2015). Towards a definition of the Internet of Things (IoT). *IEEE Internet Initiative*, 1, pp. 1–86.
29. Kranz, M., Holleis, P., and Schmidt, A. (2010). Embedded interaction: Interacting with the internet of things. *Internet Computing, IEEE*, 4(2), pp. 46–53.
30. Peng, C., Tan, X., Gao, M., and Yao, Y. (2013). Virtual reality in smart city. In *Geo-informatics in resource management and sustainable ecosystem*. Springer, pp. 107–118.
31. IEEE 2013-2019. (2019). IEEE Approved Draft Standard for an Architectural Framework for the Internet of Things (IoT). IEEE Standards Association.
32. Somayaji, S. R. K., Alazab, M., MK, M., Bucchiarone, A., Chowdhary, C. L., and Gadekallu, T. R. (2020). A framework for prediction and storage of battery life in IoT devices using DNN and blockchain. arXiv preprint arXiv:2011.01473.
33. Minerva, R., Biru, A., and Rotondi, D. (2015). Towards a definition of the internet of things (IoT). *IEEE Internet Initiative*, 1(1), pp. 1–86.
34. Darshan K. R. and Ananda Kumar K. R. (2015). A comprehensive review on usage of Internet of Things (IoT) in healthcare system. 2015 International Conference on Emerging Research in Electronics, Computer Science and Technology (ICERECT), Mandya (pp. 132–136). doi: 10.1109/ERECT.2015.7499001.
35. Samuel, S. S. I. (2016). A review of connectivity challenges in IoT-smart home, 2016 3rd MEC International Conference on Big Data and Smart City (ICBDSC), Muscat (pp. 1–4). doi: 10.1109/ICBDSC.2016.7460395.
36. Zanella, A. and Vangelista, L. (2014). Internet of things for smart cities. *IEEE Internet of Things Journal*, 1(1).
37. Atzori, L., Iera, A., and Morabito, G. (2010). The internet of things: A survey. *Computer Networks*, 54(15), pp. 2787–2805.
38. Zanella, A., Bui, N., Castellani, A., Vangelista, L., and Zorzi, M. (2014). Internet of things for smart cities. *Internet of Things Journal, IEEE*, 1(1), pp. 22–32.
39. Zhang, J., Gong, J., Lin, H., Wang, G., Huang, J., Zhu, J., Xu, B., and Teng, J., (2007, October). Design and development of distributed virtual geographic environment system based on web services, *Information Science*, 177(19), pp. 3968–3980.
40. Al-Fuqaha, A., Guizani, M., Mohammadi, M., Aledhari, M. and Ayyash, M., (2015). Internet of things: A survey on enabling technologies, protocols, and applications. *IEEE Communications Surveys & Tutorials*, 17(4), pp. 2347–2376.

41. Miraz, M. H., Ali, M., Excell, P. S. and Picking, R. (2015). A review on Internet of Things (IoT), Internet of Everything (IoE) and Internet of Nano Things (IoNT), 2015 Internet Technologies and Applications (ITA), Wrexham (pp. 219–224). doi: 10.1109/ITechA.2015.7317398.
42. Kortuem, G., Kawsar, F., Fitton, D., and Sundramoorthy, V. (2010). Smart objects as building blocks for the internet of things. *Internet Computing, IEEE, 14*(1). pp. 44–51.
43. Tayeb, S., Latifi, S. and Kim, Y. (2017). A survey on IoT communication and computation frameworks: An industrial perspective. 2017 IEEE 7th Annual Computing and Communication Workshop and Conference (CCWC), Las Vegas, NV, USA, pp. 1–6. doi: 10.1109/CCWC.2017.7868354.
44. Khoosal, D. I. and Jones, P. H. (1989). Community care again: a need for definition (pp. 451–452).
45. Atzori, L., Iera, A., and Morabito, G. (2010). The internet of things: A survey. *Computer Networks, 54*(15), pp. 2787–2805.
46. Cappelle, C., El Najjar, M. E., Charpillet, F., and Pomorski, D. (2012, May). Virtual 3d city model for navigation in urban areas. *Journal of Intelligent and Robotic Systems, 66*(3), pp. 377–399.
47. Mishra, A. K., Tripathy, A. K., Puthal, D., and Yang, L. T. (2018). Analytical model for sybil attack phases in internet of things. *IEEE Internet of Things Journal*, 6(1), pp. 379–387.
48. Alseiari, F. A. A. and Aung, Z. (2015, October). Real-time anomaly-based distributed intrusion detection systems for advanced Metering Infrastructure utilizing stream data mining. In 2015 International Conference on Smart Grid and Clean Energy Technologies (ICSGCE) (pp. 148–153). IEEE.
49. Jolliffe, I. T. and Cadima, J. (2016). Principal component analysis: a review and recent developments. *Philosophical Transactions of the Royal Society A: Mathematical, Physical and Engineering*.
50. Zheng, Y., Zhang, L., Xie, X., and Ma, W. Y. (2009, April). Mining interesting locations and travel sequences from GPS trajectories. In Proceedings of the 18th international conference on World wide web (pp. 791–800). ACM.
51. Arrington, B., Barnett, L., Rufus, R., and Esterline, A. (2016, August). Behavioral modeling intrusion detection system (bmids) using internet of things (IoT) behavior-based anomaly detection via immunity-inspired algorithms. In 2016 25th International Conference on Computer Communication and Networks (ICCCN) (pp. 1–6). IEEE.
52. Molinari, A., Maltese, V., Vaccari, L., Almi, A., and Basssi, E. (2014). *Big data and open data for a smart city*. IEEE-TN Smart Cities White Papers, Trento, Italy.
53. Gea, T., Paradells, J., Lamarca, M., and Roldan, D. (2013). Smart cities as an application of internet of things: experiences and lessons learnt in barcelona. In Innovative Mobile and Internet Services in Ubiquitous Computing (IMIS), 2013 Seventh International Conference on. (pp. 552–557). IEEE.
54. Haller, S., Karnouskos, S., and Schroth, C. (2009). *The internet of things in an enterprise context*. Springer.
55. Caragliu, A., Bo, C. D., and Nijkamp, P. (2011). Smart cities in Europe. *Journal of Urban Technology, 18*(2).

2 Green Computing – Uses and Design

Rajendran Sindhu and P. Srividya
Department of Electronics and Communication,
R. V. College of Engineering Bangalore-560059,
Email: sindhur@rvce.edu.in, Tel: +91-9538351009

1 INTRODUCTION

There is a tremendous utilization of natural energy resources by humans to meet industrial activity globally. Technological advancements have led to the use of computers for a variety of applications. This has caused an increase in global warming over the last few decades. Today, the life span of most IT products is too short. There is an excessive use of natural resources and huge e-waste due to the take, make, use, and disposal of these IT products. Hence, in a circular economy, the products should be durable, repairable, scalable, and recyclable as shown in Figure 2.1. This increases the lifetime of the products and minimizes waste. The existing IT technologies, practices, and algorithms must be redefined to obtain energy-efficient and sustainable operations. Improving the efficiency of the computer system by reducing its impact on living beings and on the environment is called green computing. It helps improvise system performance, the strength of the economy, and abides by the ethical responsibilities of humans. It includes a sustainable environment, energy-efficient, disposal cost, and recycling. In recent times, green computing has evolved as a challenge to most IT sectors that involve distributed architectures like grid, cloud, clusters, and smart grids.

Green computing deals with the following aspects:

1. Various means of using a computer in an eco-friendly way.
2. Decreasing the industrial impact on the environment.
3. Reduction in power consumption and energy.
4. The efficient use of computers leads to a reduction in environmental dissipation.
5. Decreasing the impact of advanced technology on the environment.
6. Prolong the lifetime of computer products and improvise the efficiency of producing energy.
7. Usage of recyclable products in computer manufacturing.
8. Usage of lesser hazardous and biodegradable materials to reduce environmental pollution.
9. Reduction of industrial wastes during energy production.

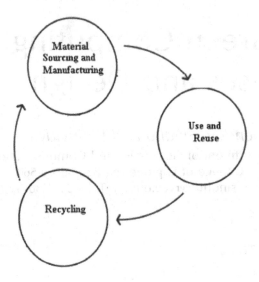

FIGURE 2.1 Circular Economy to Reduce Wastage.

10. Providing alternatives to technologies that harm the environment.
11. Proper disposal of e-waste.

Green computing is inclusive of both hardware and software technologies.

Software technologies: This includes various software aspects like enhancing the coding techniques to improve program efficiency and reduce storage requirements. It also includes various computing models like distributed computing, cloud computing, mobile computing, IoT, and others.

Hardware technologies: This includes modification of computer hardware like computers, servers, printers, projectors, and other digital devices so it helps reduce energy consumption and greenhouse gas emission. Recycling and waste management are highly encouraged under this.

2 ORIGIN OF GREEN COMPUTING

The concept of sustainable development became popular when the World Commission on Environment and Development released a report called "Our Common Future" in 1987 [1]. With this, social activists and other environmentalists started working and, as a result, a consumer energy plan was launched by the US Environmental Protection agency in 1992. The main goal was to bring down energy consumption and reduce the emission of greenhouse gases. At the start, this plan concentrated on computer program and was later extended to all sectors.

Currently, there are various organizations that develop green computing standards. These organizations insist that all IT industries use certified products that meet environmental and social benchmark throughout the product's life cycle. The benchmarks laid are on energy efficiency, usage of hazardous substances, and ergonomic design. Some of the standards developed by various groups are listed below:

1. Video Electronics Standard Association (VESA) is an organization that sets, promotes, and supports interface standards for computer displays, workstations, and computing environments. It was established in July 1989 in California. It sets standards related to power consumption of display units, display resolution, display interface compression standards, and others. The first standard developed was the 800 × 600 Super Video Graphics Array (SVGA) and its software interface.
2. The Swedish confederation of professional employees developed TCO certification. It is the widest sustainable certification for all IT-related products. It helps develop the most responsible product that aids in the sustainable direction of the industry. This certification also supports the organization's efforts to reduce the risk and take the next step towards environmental and social responsibilities. TCO certification is available for various IT products like projectors, notebooks, tablets, desktops, and data center products like data storage servers and network equipment.
3. American National Standards Institute (ANSI) is a non-profit private organization in the United States that supervises the development and usage of standards by accrediting the processes of the organizations that develop them; it doesn't develop any standards on its own. The institute also synchronizes US standards with other international standards to enable the use of American products worldwide. ANSI accreditation implies that the processes of the organizations meet all the institute's requirements in terms of environmental constraints, consensus, and openness. It adopts the Electronic Product Environmental Assessment Tool (EPEAT) system to certify electronic products that impact the environment. EPEAT is a method that assesses various environmental aspects of a product and—based on certain environmental criteria—classifies the product into three categories: gold, silver, or bronze.
4. Green Electronics Council is an organization that manages EPEAT. In this system, the manufacturers should declare the environmental criteria met by their product. This will be verified by an assurance body and declared whether the product meets the IEEE 1680 Green Electronics Standards.

3 RELATED WORKS

At present, almost all the products (e.g., smart devices, wireless earphones) are comprised of an embedded processor. Although the processor offers higher performance capability to the device, it consumes a lot of energy for computations. A novel strategy using Hybrid Asymmetric Multicore Architecture (HAMA) and parallelism is proposed by Hope Mogale et al. in [2] to reduce the energy consumed during computations and accelerate green computing. HAMA uses a self-timing mechanism that ensures that the cores are powered off when not in use. To avoid computational overheads, structural parallelism is adopted.

Global warming has increased due to excessive trimming down of trees to meet the demand for paper. As a consequence, there is an excessive increase in the earth's temperature and carbon content in the atmosphere. The research [3] suggests a paperless approach using modern technology to minimize waste.

Research in the area of mobile cloud computing involves parallel and distributed systems comprised of interconnected mobile devices and computing clusters. Construction of the actual computing cluster, its scalability, and efficiency are discussed in [4].

The amount of carbon dioxide emitted by cloud computing servers is increasing vastly, leading to pollution. Paper [5] discusses the six different scheduling algorithms for green computing. It is a two-step process: assigning maximum possible tasks with lower energy to a cloud server and assigning the same best speed for all tasks scheduled to the server. All six algorithms were simulated and the shortest task first algorithm was the best fit for green computing.

4 NEED FOR GREEN COMPUTING

With the fast-growing economy, business is implemented across every sector available. Companies need to analyze, store, track, and collect huge volumes of information and data logs to mobile call records, which requires a lot of investment and data warehouses—thereby consuming a huge amount of power and large servers yielding around 61B kWh of electricity, at a cost of $4.5 billion annually. IT industries have particularly addressed the energy consumption in data centers by incorporating various approaches. As data increases, the complexity of the hardware is an issue in all the approaches; so far, this can be negated by using tightly-packed servers. To reduce their hardware footprint, industries need to shrink their "data footprint" by addressing how much server space and resources their information analysis requires.

A combination of new database technologies explicitly designed for analysis of massive quantities of data and affordable, resource-efficient, open-source software can help organizations save money and become greener. Organizations can do so in the following key areas: reduced data footprint, reduced deployment resources, and reduced ongoing management and maintenance. The technology has a lot of benefits—energy saving during idle operation, energy consumption during the peak time of computing resources, eco-friendly energy, and reduction of computing waste. Harmful effects of computing resources and the problems of reducing environmental effects from fossil fuel emission have raised to the top of global public policy. To embrace environmentally-sustainable products, industries and businesses have offered solutions using low carbon that reduces cost and is energy-efficient, thereby reducing global greenhouse gas emissions.

Green Computing – Uses and Design

5 CHALLENGES IN GREEN COMPUTING

In any technology, there will be challenges. Similarly, there are many challenges in green computing such as disposal of electronic waste, investments, energy efficiency, etc.

5.1 Return of Investment

Not many stakeholders are aware of the environmental impact of computers since we will not be able to demonstrate immediate results, which is the disadvantage of the project. The returns for these projects are expected for a long time. The most challenging issue is to showcase the instant results after the implementation of Green IT. In this regard, there are effective methods for the replacement of costly and heavy devices such as processors, printers, etc. Environmental-friendly disposal of the system had to be worked on as the current infrastructure has inefficiencies and it is vital to know the environmental impact of computers.

5.2 Disposal of Electronic Wastes

One of the biggest challenges faced by the electronics industry is the reliability of the use of green materials in computers. The focus is to bring an eco-friendly range of computers so that the e-waste in the environment is reduced. Hazardous materials such as brominates, flame-retardants, heavy metals, and PVCs are used in manufacturing computers, which can be replaced by metals that have high melting temperatures such as silver/copper/tin alloy.

5.3 Perspective with Respect to Indian Scenario

Due to socio-economic matters, India is facing a dilemma for the implantation of green computing. The lack of research initiative and congenial infrastructure has resulted in the absence of indigenous commercial products and good patents. In the past few years, tax relief given by the government has accelerated the import of computer hardware, resulting in the minimization of machines, peripherals, and equipment. In this state, many small- and medium-scale industries were induced for procuring hardware at low prices and endeavor into the building of IT infrastructure for the company.

5.4 Energy Efficiency and Miniaturization Board Level Witnessing the R&D Capabilities

Energy efficiency in integrated circuit chip design requires focus on reducing voltage while holding performance, enabling integration within smaller packages, which is a tedious process that requires highly-skilled engineering—to offer low power consuming embedded, notebook, and processors to the world. Smaller systems with a high level of energy efficiency and low heat production provide savings for consumers in terms of disposal costs and power requirements. For portable structures, such as the new era of extremely-cellular gadgets, this additional method of longer battery lifestyles requires a vital detail of ultra-mobility. Other groups cannot acquire this, as their chips devour too much power and emit an excessive amount of heat.

6 PARADIGMS OF GREEN COMPUTING IN IT

Green computing in IT focuses on the following paradigms:

1. Green design
2. Green manufacturing
3. Green use
4. Green disposal

Green design: It deals with designing computers, printers, scanners, projectors, servers, and other IT products that are energy-efficient and environmental-friendly. It recommends the use of materials that will not go to waste when the products are recycled. The components involved in green design are as shown in Figure 2.2.

Product Longevity: It aims to develop a long-lasting product. This reduces manufacturing efforts to some extent.

Modularity: It is a measure that shows how much a system can be broken down into smaller parts. This simplifies the repair of parts and system upgrades but requires interfaces to connect the parts.

Components Size: This is a step towards a green initiative. The smaller the component size, the lesser the material required to build it, and the lesser is the energy required to function.

Packaging: This is usually done using renewable or recyclable materials to reduce the amount of resources and waste.

End of Life Cycle: This should be planned meticulously to maximize the parts of the system at the end of the product life cycle.

Green manufacturing: It deals with materials used to manufacture various components of the computer with energy-saving technologies that can be adopted to make computer parts. For example, the materials selected should be recyclable and environment-friendly. Components like Light Emitting Diode monitors can be used to reduce energy consumption.

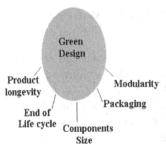

FIGURE 2.2 Components Involved in Green Design.

Green Computing – Uses and Design 25

Green use: The computer and its related products should be used in a way that the energy consumption and greenhouse gas liberation are minimized. Some of the steps towards this effort include the following:

a. Changing the work habit of the computer user. For example: turning off the computer when not in use for more than five minutes will save a lot of energy. The use of notebook computers instead of personal computers also helps save energy. Avoiding screensavers will also save energy.
b. Telecommuting: Organizations allowing employees to work from home by providing all the facilities required for their work is termed as telecommuting. This helps reduce fuel consumption required for their commute.
c. Virtualization: This refers to the consolidation of servers. It allows more than two computer systems to run on the same physical hardware. This helps reduce power consumption and also leads to cooling.
d. Voice Over Internet Protocol (VOIP): In this technology, the internet or any other packet-switched network will be used to transmit voice. This helps reduce telephone wiring, further reducing metallic waste, wiring cost, and wiring infrastructure.

Green disposal: Saving the environment requires the reuse of old computers and the saving of unwanted products.

7 APPLICATION OF GREEN COMPUTING

Green Computing can be used in various applications, as explained below:

7.1 IN DISTRIBUTED AND CLOUD ENVIRONMENTS

Distributed computing is an architecture where software system components are shared by multiple computer systems within a narrow geographical area. Distributed computing involves a client-server model for communication. This helps improve efficiency and performance. This is achieved by distributing various processes in business at efficient places over the entire computer network. For example, in a three-tier distribution model, processing of user interface is done in the PC located at the user's location, business processing is carried out on a remote PC, and database processing and access are carried out on a PC that offers centralized access for business processes.

Today, the view of distributed computing has reshaped the field of dispersed frameworks and, generally, changed how organizations use figuring. Distributed computing is offering utilities arranged IT administrations to clients around the world. It empowers the facilitating of uses from purchaser, logical, and business spaces. Server farm's facilitating distributed-computing applications devour gigantic measures of vitality, adding to high operational expenses and carbon impressions to the earth. With vitality deficiencies and worldwide environmental change driving our interests nowadays, the force utilization of server farms has become a key issue. Green figuring is additionally getting progressively significant in a world with restricted vitality assets and an ever-rising interest for increasingly computational force. We need green-distributed computing arrangements that can spare vitality and additionally decrease operational expenses; structural systems and rules that give productive green upgrades inside a versatile Cloud registering engineering with asset provisioning and distribution calculation for proficient administration of distributed computing situations, improving vitality effectiveness of the server farm. Utilizing power-mindful planning strategies, variable assets of the board, live relocation, and an insignificant virtual machine structure, framework proficiency will be boundlessly improved in a server farm-based Cloud with negligible execution overhead.

For multiple IT services, cloud computing has become an enabling technology. Cloud computing is an architecture where customers make use of servers for storage and pay for it based on utilization. Since the availability of resources is variable, cloud computing is dynamic in nature. The resources are allocated to the customers through the Cloud Data Center (CDC). The data centers consist of several servers, storage devices, and network devices in the same complex. The servers are also computers without peripherals (monitor, keyboard, and mouse). It only acts as a storage device and is connected to the network to make the stored data available to other computers connected to the network. The CDC provides a wide range of services ranging from high-performance computing to data analytics on large scale up to end-users.

According to worldwide statistics, the CDC that are set up at various geographical locations contributes to about 25% of the total share of electricity consumption by IT sectors. The servers in the CDC also get heated quickly and require a cooling mechanism that further consumes energy. CDC also results in an enormous amount of carbon emission called greenhouse gases. It leads to increased disease rates, ozone layer depletion, and global warming.

Hence, proper strategies are required to reduce energy consumption and carbon dioxide emission to protect the environment. The main aim of green computing is to utilize computing resources efficiently to reduce the energy consumption by computer systems, improvise recycling capabilities, and reduce emissions to prevent environmental degradation and make the planet cleaner and greener.

As per the data issued by Greenpeace, a non-government and environmental organization, the electricity consumed by these data centers globally amount to 32 GW—which is the power consumption of more than 1.5 lakh houses.

Techniques that can be adopted to obtain green CDC are:

1. Efficient resource management;
2. Utilization of heat generated;
3. Usage of renewable energy;
4. Task scheduling using efficient algorithms;
5. Green cloud computing using Genetic Algorithm; and
6. Pre-emptive priority-based job scheduling algorithms.

1. **Efficient resource management:** Efficient use of resources can be done by the virtualization process. Virtualization in CDC provides scalable and fault-

resistive operations. The virtualization layer lies on the physical resources and abstracts the physical resources layer to provide user interface and applications. The virtualization layer manages CDC resources through various backup techniques like snapshots and resource migration. Virtualization, thus, helps in resource consolidation and higher energy efficiency.

2. **Utilization of heat generated:** The heat that is generated from the CDC can be reused for cooling purposes. The heat dissipated from the servers is sent through cooling systems, where thermal energy is pumped to cooler space by reversible heat pumps. The CDC itself can be located in cooler regions to reduce heat generation.

3. **Usage of renewable energy:** The use of renewable energy reduces carbon emission. Solar or wind energy can be used as renewable energy in the CDC, but the main drawback of renewable energy sources is high installation cost and unpredictable supply availability. The problem of unpredictable supply availability can be resolved by using few techniques like dynamic power balancing, server power capping, and using hybrid power supplies [6].

4. **Task scheduling using efficient algorithms:** Task scheduling is an important aspect of cloud computing. It refers to the allocation of tasks to the best resource for execution. Numerous techniques like integer programming, graph theory, genetic algorithms, and others are proposed in the literature to obtain an optimal solution for task scheduling of resources by the data centers.

   Among the various techniques available, genetic algorithms proved to be the best by offering uniform distribution of resources across the globe to all customers. This is referred to as load balancing.

   Load balancing ensures that excess dynamic workload gets evenly distributed across the entire network and leads to efficient resource utilization, bottleneck avoidance, reduced response time, customer satisfaction, and reduced energy dissipation. Reduction in energy consumption is achieved due to a reduction in overheat and carbon emissions [7]. This load balancing helps procure a green environment over cloud computing. Different options for green computing in the cloud are as shown in Figure 2.3.

5. **Green Cloud Computing Using Genetic Algorithm:** Cloud computing provides the customers access to the required resources anytime from any location, through the internet. However, Internet Data Centers (IDCs) demand more energy and are the main cause of CO_2 emission. Demand for resources may be CPU-intensive or I/O-intensive. These requirements are to be met by cloud resources to maintain Quality of Service using SLAs. Green cloud computing can be achieved using a Genetic algorithm for load balancing. This is achieved by using individual server metrics like thermal design points to obtain a balanced distributed load vector. The distribution of workload ensures a reduction in energy consumption and lowers carbon footprint. Cloud computing conveys a capable method to proficiently finish their administration requests. A versatile virtual machine scheduling algorithm was proposed in the paper to deal with the issues of VM scheduling in situations pertaining to cloud computing. Improvement in the performance is a need of the hour in

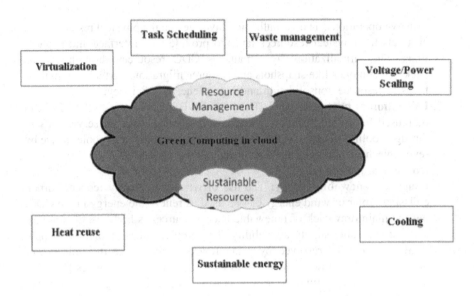

FIGURE 2.3 Options for Green Computing in Cloud [8].

the development of computer systems because of the demand from consumers. To effectively utilize the resources for storage and server's green cloud computing approach is used to consume less energy. Carbon dioxide emission is due to the intensive use of energy, especially from disposable energy, the resources are available by means of internet data centers. The utilization of virtual machine designation calculations for distributed computing system furnishes extraordinary flexibility with the ability to relocate virtual machines across physical machines

Genetic algorithms are adaptive search algorithm based on genetics and natural selection, which is rapidly growing. The algorithm impersonates the process of evolution based on the population of candidate solutions. It is best suited for complicated or NP-hard problems because of its simplicity and robustness. In the process of evolution, a modification is performed by using genetic operators on each individual. Each chromosome represents a load balancing result, and an evaluation fitness function is called to evaluate the offspring.

6. **Pre-emptive Priority Based Job Scheduling Algorithms:** Green Cloud is a bundle test system that centers around boosting the framework, with sparing vitality on various servers. In Green Cloud Computing, Occupation Scheduling is one of the significant issues. To deliver the greatest throughput of Green Clouds, the work must be finished by organizing employments on each cloud. Another system is the Pre-emptive Priority Based Job Scheduling Algorithm in Green Cloud Computing (PPJSGC)—the pre-emptive part ascertains the vitality utilization for planning the occupations on the processing

servers. The registering servers are assigned to forms dependent on the best fit according to their vitality necessities and server recurrence accessibility. Dynamic Voltage and Frequency Scaling (DVFS) Controller is used to calculate performance in this activity. The heap of executives, low vitality utilization, and boost in the income are the key thought process of our investigation [9–12].

Server rooms

Most organizations don't maintain data centers; all their IT equipments are stored in a server room that will act as a data center at a smaller level.

Some of the measures that can be adapted to maximize server room efficiency are:

1. Usage of large servers
2. Raised flooring in server rooms and computer rooms
3. Usage of variable speed motors
4. Racks with inbuilt chillers or water-cooled servers can be used
5. Replacing physical servers instead of virtual servers

Therefore racks (19-inch EIA-310-D standard) with built-in chillers or even water-cooled servers, become preferable.

7.2 IN SMART PORTABLE DEVICES

Long term evolutionary (LTE) communications have enabled smart portable devices like tablets and phones to have powerful communication and computing abilities to perform resource-intensive tasks. Smartphones have reduced the dependency of users on laptops and desktops to perform certain tasks. This has increased the demand for smartphone resources. Smartphone applications make use of GPS, wireless radios, accelerometers, and other services to meet user requirements.

The higher data rate services also causes more energy to drain out of the device. The existing phone batteries are not that efficient to retain energy even for a full day whenever the device is used for large volumes of data transfer or video purposes. Hence, more focus should be done to develop energy-efficient, high-computing portable devices. Green computing techniques can be applied to upgrade the network's performance by developing energy-efficient devices.

The energy consumption of smartphones can be reduced by efficient management of hardware components of the device. As Complementary Metal Oxide Semiconductor (CMOS) occupies a very small data space, it is preferred in manufacturing an IC. Static, dynamic, and leakage powers are the main causes of power consumption in CMOS-based circuits. Static power is the amount of power consumed when there is no transition in the state; dynamic power is the power consumed when there is a transition in the state (either from logic 0 to logic 1, or vice versa). Leakage power can be reduced by using a high voltage threshold (high VT) transistor between the circuit ground and the actual ground of the device. Software-based solutions like mobile cloud computing based on computational offloading, energy-efficient code designing, and handling the energy bugs can also be provided to reduce the energy consumption of the device.

Mobile cloud computing allows smartphones to enhance the lifetime of the device by offloading the tasks, requiring more energy on cloud servers. Computational offloading considers resource utilization, energy requirements, total time for execution, and other privacy concerns of an application and migrates the tasks to cloud servers.

Handling the energy bugs: Energy bugs are difficult to identify and leads to unusual power consumption in smartphones. The main cause of energy bugs is the use of damaged smartphone chargers, defective batteries, defective SIM cards, defective memory cards, etc. Usage of these defective gadgets also drains energy.

Energy-efficient code designing: Inefficient codes have a huge impact on the energy consumption of smartphones. Infected mobile application also drains the smartphone application drastically. Some code optimizations that can be adopted are listed below.

i. The best placement of functions and classes helps reduce energy consumption. Example memory distance between two frequently communicating functions reduces energy consumption.
ii. Energy-efficient techniques like loop unrolling, branch predictions, instruction pipelining, and instruction scheduling can be used.

The various hardware and software options to achieve green mobile computing is shown in Figure 2.4.

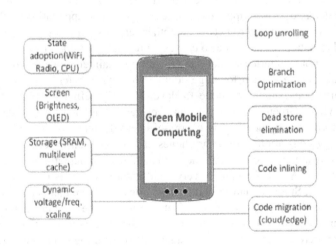

FIGURE 2.4 Techniques for Green Mobile Computing [8].

Green Computing – Uses and Design

TABLE 2.1
Various IoT Enablers and Greening Strategies

IoT Enabler	Type	Data Transfer	Power Source	Life-Time	Greening Strategies
RFID	Active tags	Low	Battery	≤ 5	Energy-efficient algorithms and protocols
	Passive tags	Very low	Harvest	∞	
Sensing network	Smart object	Low	Battery	≤ 5	Sleep wake-up, data reduction mechanisms
	Mobile sensing	High	Battery	≤ 2	
Internet technologies	Cloud	Very low	Grid	≤ 10yrs	Turn off unrequired facilities, Minimize data path length
	Future internet	Very low	Grid	∞	

7.3 IN IoT

IoT is a technology that allows data communication among various electronic devices without human or computer intervention. Green IoT is achieved by adopting a set of efficient hardware and software techniques. The main aim of Green IoT is reducing carbon emissions and achieving a cleaner environment. Various IoT enablers and greening strategies are shown in Table 2.1.

Green RFID: Radio frequency identification (RFID) is a prominent enabler of IoT. It consists of an RFID tag and a tag reader. RFID tag is a radio with an attached microchip—it acts as a transceiver. Every tag has a unique ID that stores data relevant to the entities it is attached to. RFID tag reader transmits a signal and triggers the flow of information, and then receives the responses that come from the nearby tags. The range of communication is usually less than a few meters.

Green RFID can be achieved using the following methods:

1. Since the recycling of tags is difficult, RFID tag size should be decreased.
2. RFID tags can be manufactured using biodegradable materials.
3. Using energy-efficient algorithms and protocols.

Green wireless sensor networks: Recent advances in system administration, storage, and registration profoundly affect the improvement of cutting-edge green remote systems. These three significant zones have generally been tended in existing works for an epic structure that considers organizing, reserving, and figuring strategies in a precise method to normally bolster vitality effective data recovery and registering administrations in green remote systems. This incorporated system can empower a dynamic organization of various assets to meet the prerequisites of cutting-edge green remote systems.

A WSN consists of abundant sensor nodes with resource constraints like limited computing, storage capacity, and limited power. The sensors are connected to a base station called a sink. The sensor nodes can read the surrounding environmental conditions like temperature, humidity, etc.

Green WSN can be achieved using the following methods:

1. Activating sleep mode when the sensor is idle
2. Energy-efficient routing
3. Radio optimization

7.4 IN PARALLEL COMPUTING OF BIG DATA SYSTEMS

Big Data is ordinarily sorted out around a dispersed record framework where equal calculations can be executed to understand Big Data investigation. The equal calculations can be mapped in various elective manners to the processing stage. Therefore, every elective will perform contrastingly regarding the earth's significant parameters—for example, vitality and force utilization. Existing examinations on the sending of equal figuring calculations have, for the most part, centered around tending to general registering measurements (e.g., speedup as for sequential processing and effectiveness of the utilization of the registering hubs). To examine sending options, the elicitation of green measurements for an enormous information framework is required. The significant green registering measurements for enormous information frameworks are adapted [8].

8 PROS AND CONS OF GREEN COMPUTING

Green computing (in converting authorities policy) inspire recycling and reducing power use with the aid of industries and people. The inexperienced computing strategies make use of less strength because it interprets with lower carbon dioxide emissions, thereby decreasing fossil fuel that can be used in strength flow and transportation. Due to this less power needed to provide, dispose, and use products, conserving strength and assets save a lot of money. The most crucial advantage is decreasing the hazard in laptops along with chemicals that cause nerve damage, cancer, and immune reactions in human beings.

Because of the rapid exchange in technology, some inexperienced computer systems may be considered underpowered, making it costly [3].

9 MEASURES TO BE TAKEN IN DEVELOPING IT PRODUCTS AS A MOVE TOWARDS GREEN COMPUTING

a. Socially and environmentally responsible manufacturing—excluding environment conflict materials in product development, a robust design to reduce the energy consumption by the products, developing a product with low noise levels, low electric and magnetic fields, and a product that is electrically safer.
b. Extending the product lifetime by developing a durable product that can withstand temperature variations, with extended lifetime and replaceable batteries that use standard connectors.
c. Usage of recyclable materials.

10 CONCLUSION

With the advent of emerging technologies—cloud computing, mobile computing, big data analytics, and IoT (i.e. the driving forces in the industries)—there have been several algorithms and protocols for the green systems in the industry. In this chapter, we have discussed the need for green computing, its evolution in the industry, and the different challenges faced by green technologies. We have also emphasized green computing, its applications, and its advantages and disadvantages.

REFERENCES

1. Anthony, B., Majid, M. A., and Awanis, R. (2017). A descriptive study towards green computing practice application for data centers in IT based industries. MATEC Web of Conferences Volume 150, 2018 Malaysia Technical Universities Conference on Engineering and Technology (MUCET).
2. Mogale, H., Esiefarienrhe, M., Gasela, N., and Letlonkane, L. (2019). Accelerating green computing with hybrid asymmetric multicore architectures and safe parallelism. In 2019 International Conference on Advances in Big Data, Computing and Data Communication Systems (pp. 1–7). Winterton, South Africa.
3. Al Sadoon, G. M. W., Makki, H. A., and Saleh, A. R. (2017). Green computing system, health and secure environment management system. In 2017 4th IEEE International Conference on Engineering Technologies and Applied Sciences (ICETAS) (pp. 1–6). Salmabad.
4. Kalange Pooja R. (2013). Applications of green cloud computing in energy efficiency and environmental sustainability. *IOSR Journal of Computer Engineering (IOSR-JCE)*, pp. 25–33. ISSN:2278-0661, ISBN:2278-8727.
5. Zhang, L. M., Li, K., and Zhang, Y. (2010). Green task scheduling algorithms with speeds optimization on heterogeneous Cloud servers. In 2010 IEEE/ACM International Conference on Green Computing and Communications and International Conference on Cyber, Physical and Social Computing (pp. 76–80). Hangzhou.
6. Yu, X., Ma, Y., and Li, J. (2018). Analysis and research on green cloud computing. In 2018 2nd IEEE Advanced Information Management, Communicates, Electronic and Automation Control Conference (IMCEC) (pp. 521–524). Xi'an.
7. Tripathy, A. K., Das, T. K. and Chowdhary, C. L. (2020). Monitoring quality of tap water in cities using IoT. In *Emerging Technologies for Agriculture and Environment* (pp. 107–113). Springer, Singapore.
8. Zhang, X., Gong, L., and Li, J. (2012). Research on green computing evaluation system and method. In 7th IEEE Conference on Industrial Electronics and Applications (ICIEA) (pp. 1177–1182).
9. Zhang, X., Gong, L., and Li, J. (2012). Research on green computing evaluation system and method. In 2012 7th IEEE Conference on Industrial Electronics and Applications (ICIEA). (pp. 1177–1182). Singapore.
10. Huo, R. et al. (2016, November). Software defined networking, caching, and computing for green wireless networks. *IEEE Communications Magazine*, 54(11), pp. 185–193.
11. Gürbüz, H. G. and Tekinerdogan, B. (2016). Software metrics for green parallel computing of big data systems. In 2016 IEEE International Congress on Big Data (BigData Congress) (pp. 345–348). San Francisco, CA.
12. Al Sadoon, G. M. W., Makki, H. A., and Saleh, A. R. (2017). Green computing system, health and secure environment management system. In 2017 4th IEEE International Conference on Engineering Technologies and Applied Sciences (ICETAS) (pp. 1–6). Salmabad.

Part II

Analysis

3 Statistical Methods for Reproducible Data Analysis

Sambit Kumar Mishra, Mehul Pradhan, and Rani Aiswarya Pattnaik

Department of Computer Science and Engineering, Siksha 'O' Anusandhan Deemed to be University, Bhubaneswar, Odisha, India

1 INTRODUCTION

This chapter deals with the methodologies and the process of data processing and how it is helpful to other domains. The process is comprised of methods treating outliers and missing values, processing the data, preparing the respective model, and showcasing the model using proper statistical analysis diagrams [1,2]. This chapter explores some machine learning topics: inferential statistics, predictive modeling, and supervised and unsupervised learning. Here, the detailed report on various approaches dealing with both structured and unstructured data is presented [3]. The language used for data science is Python. In this chapter, there may exist some algorithms in R language also. There are several types of libraries used, such as pandas, NumPy, etc. that will help initiate the methods. There will also be some histograms, graphs, and tables in the complete chapter for a better interpretation of the concepts that will be explained. This chapter will touch on all necessary and known methodologies of data handling while briefing about the machine learning world [4].

The data science overall helps manage various parts of an organization. It is the backbone of the data with which a company must deal with regularly. It somehow helps integrate every possible way to deal with data [5,6]. Data manipulation widely helps in business and financial modeling — it studies the present data, manipulates it using proper tools, and predicts the ways to deal with upcoming challenges. Therefore, data science is becoming an integral part of the working class and is reaching out to every corner where a problem exists. It is easing out the difficulties which lie within every task.

2 OVERVIEW HIERARCHY

We all know that we are creating and storing data at a humongous scale and can run computations on it at very low costs. This is all because of the implementations of data analytics. As per research, the data generated from the last two years is way greater than the data generated prior. This shows how important data handling has become. There exists a certain spectrum in the world of data analytics with increasingly complex levels [7].

As shown in Figure 3.1, it includes the relation between complexities and attributes of various fields in an incrementing order.

The hierarchy levels shown above are described below.

MIS (Management Information System) refers to the business profitability scored and the bad quality in the process. Detective Analysis is the domain that deals with questions like, "Why is the business in Delhi worse than the one in Mumbai?" In short, the questions derived from the MIS table are detective analysis. Dashboarding involves business intelligence created in real-time [8]. It is used to explain happenings in the business diagrammatically and statistically (e.g., pie charts, etc.). Predictive Modeling happens when a person collects all the data to know what is happening at a granular level — it basically answers the question of, "What is likely to happen at the granular level?" Finally, Big Data tells what *can* happen, given that a user collects and use all possible data.

3 INTRODUCTION TO DESCRIPTIVE STATISTICS

Descriptive statistics are descriptive coefficients that summarize a given data set that can either be a representation of the sample or a population.

Note: In the programming codes given below, Python language shall be used. To perform operations, various libraries shall be included, which will be described later as necessary. For now, we shall include the pandas library using the code (Program 1.1).

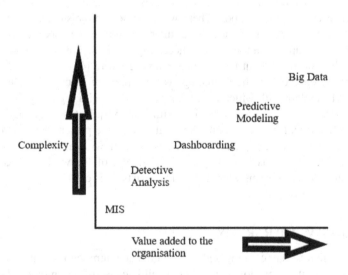

FIGURE 3.1 Spectrum of Business Analytics.

Statistical Methods

Program 1.1
```
import pandas as pd
```

Pandas is a software library that is usually used in Python programming. It offers data structures and operations for data manipulation and analysis [9].

NumPy package provides various mathematical functions, linear algebra routines, random number generator, and more. As our chapter consists of various statistical operations, the NumPy package will be helpful.

There is a need to import data set in every operation. For the record, this shall be named as "excel_filename". To import the required data set, the following code shall be used (Program 1.2).

Program 1.2
```
data=pd.read_csv("excel_filename")
```

1. **Mode**: A number that occurs most frequently in the data series [10]. It is a robust method and is not generally affected by the addition of a couple of new values. To perform mode operation, refer to Program 1.3.

Program 1.3
```
mode_data=data['column_name'].mode()
print(mode_data)
```

2. **Mean:** It is often referred to as average. The formula of mean is $\Sigma x/n$, where Σx is the sum of the quantities and n is the number of quantities. Refer to Program 1.4.

Program 1.4
```
mean_data=data['column_name'].mean()
print(mean_data)
```

3. **Median:** The median is the middle number from a sorted (ascending or descending) data set calculated. It depends on the number of data available [11,12]. In Python, this works in a very simple manner through an inbuilt function. Refer to Program 1.5.

Program 1.5
```
median_data=data['column_name'].median()
print(median_data)
```

4. **Interquartile range:** The difference between the third quadrant and the first quadrant in a box plot for a set of data is known as the interquartile range (Program 1.6).

Program 1.6
```
q1=data['column_name'].quantile(0.25)
q2=data['column_name'].quantile(0.50)
q3=data['column_name'].quantile(0.75)
q4=data['column_name'].quantile(1)
iqr=q3-q1
print(iqr)
```

5. **Histogram**: Histograms are used for continuous data by putting bins to the numerical data. A person gets an idea of the distribution from the histogram [13]. The height of the bar denotes the frequency. Some of the conditions required are:
    i. The bins should be of equal size.
    ii. The bins should not overlap.

To draw a histogram chart for a given set of data, refer to Program 1.7. Use matplot library to successfully run the given code below.

Program 1.7
```
plt.hist(x='column_name',data=histogram)
plt.show()
```

6. **Outliers**: Any values that fall outside the range of data is termed as an outlier [14]. Reasons for Outliers are:
    i. Typos – Outliers during data collection. For example, adding an extra zero by mistake.
    ii. Measurement error – Outliers in data due to measurement operator being faulty.
    iii. Intentional error – These are errors that are induced by people intentionally. For example, teens might claim they've had less amount alcohol than they actually have.
    iv. Legit Outliers – These are values that are not actually errors but are in the data due to legitimate reasons. For example, a CEO's salary might be remarkably high compared to other employers.

Outliers can be detected using a Box plot. When there is a presence of some data values outside the whiskers of the Box plot, one can observe outliers easily.

If a value exists below Q1 – 1.5 times IQR* or above Q3 + 1.5 times IQR*, the value is certainly to be an outlier.

* IQR = Interquartile Range

Statistical Methods

7. **Spread of data**: The spread of data describes how similar or varied the set of observations are. **Range** is the difference between the smallest and largest data. The bigger the range, the more spread out the data is (Program 1.8).

Program 1.8
```
max_data=data['column_name'].max()
min_data=data['column_name'].min()
range_data=max_data-min_data
print(range_data)
```

8. **Variance**: It is the average squared deviation from the mean. The most important thing to notice is that if we add a constant to each value in the data set, it does not change the distance between values, so the variance remains the same.

$$\sigma^2 = \frac{\Sigma(x - \mu)^2}{n}$$

(Formula)
where x = individual data
μ = mean of data set
n = number of items

Program 1.9
```
mean= data['column_name'].mean()
difference=data['column_name']-mean
sq_difference=difference**2
variance=sq_difference.mean()
print(variance)
```

TABLE 3.1
Difference between Range and Interquartile Range

Range	Interquartile Range
Susceptible to outliers	Robust to outliers
Range = Max-Min	IQR = Q3 − Q1
Takes entire data set into consideration	Takes 50% data set into consideration

To apply variance in programming, refer to Program 1.9.
There exists a direct function that is completely devoted to the variance.

Program 1.10
```
variance1=data['column_name'].var(ddof=0)
print(variance1)
```

(Program 1.10)

9. **Standard Deviation:** It is the square root of variance [15]. It also doesn't get affected by the change of values in the data set. Its formula is:

$$\text{Standard deviation} = \sqrt{\frac{\Sigma((x - \mu))^2}{n}}$$

Program 1.11
```
std=data['column_name'].std(ddof=0)
print(std)
```

To apply standard deviation, use Program 1.11.

10. **Frequency Table:** It is a schematic diagram that shows the data values of the data set and their frequencies in an ascending or descending manner. To apply, use Program snippet 1.12.

Program code 1.12
```
freq_data=data['column_name'].value_counts()
print(freq_data)
```

---

11. **Variables:** Memory that holds a certain kind of data.

   Types of variables with respect to the range of values:
   i. Continuous variable: These variables are those that contain values of the continuous range (e.g., 80.5–90).
   ii. Categorical variable: These contain two types of variables:
      a. Nominal Variables: Variables with no order and preference (e.g., male and female).
      b. Ordinal Variable: Variables having certain order and preference (e.g., bad, good, excellent).
   iii. **Random variable**: Each value of a random variable may or may not be equally likely [16]. If a nonfighter starts fighting a wrestling champion, the chances of winning are not equally likely. Sometimes, the random variable can only take fixed or discrete values (e.g., the sum of values of two

dice can either be 2 or 3, but not 2.5).
Y = {2,3,4,5,6,7,8,9,10,11,12} Total outcome = 36
Binary outcomes: There are always two outcomes in a binary outcome problem. For example, head or tail, win or loss, etc. There is either a 50-50 probability or none. For example, heads or tails have a 50-50 probability; winning and losing may have 70-30 or 20-80, or something else.

iv. **Bernoulli trials**: An experiment that has exactly two outcomes.
v. **Binomial Distribution**: The probability distribution of the number of success in n Bernoulli trials is known as a binomial distribution.

$$\text{Formula} = {}^nC_k \times p^k \times q^{n-k}$$

vi. **Continuous Random Variable**: These variables can take any values in a given range. For example, the amount of sugar in orange: 2.4 g, 2.45, or 2.456 g.

Central Limit theorem: Central limit theorem establishes that, in some situations where independent random variables are added, their properly normalized sum tends toward a normal distribution (informally a bell curve) even if the original variables themselves are not normally distributed [17] (Figures 3.2–3.4).

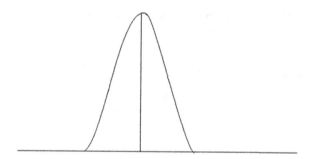

FIGURE 3.2 Normal Distribution.

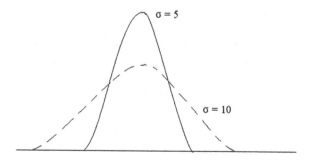

FIGURE 3.3 Result in Change of Standard Deviation.

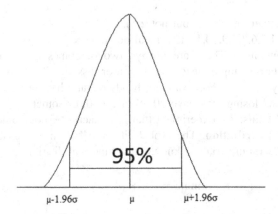

FIGURE 3.4 Distribution of C.I.

Properties of Normal Distribution:

i. The distribution is symmetric near the mean.

As in Figure 3.2, one can observe how the shape is symmetrical by the mean.

i. The higher the standard deviation, the greater number of items that deviated from the mean. Refer to Figure 3.3.

4 INTRODUCTION TO INFERENTIAL STATISTICS

This includes phenomenon such as inferences about the population from the sample. This helps know whether a sample is significantly different from the population [18].

Some Terminologies:

i. Statistic – A single measure of some attributes of a sample.
ii. Population statistic – The statistic of either population in context.
iii. Sample statistic – The statistics and the analysis results of a sample taken from a population.

1. **Z-score**: The distance in terms of structural deviations; the observed value from the mean is the standard score of z-score. Its observed value is $\mu + z\sigma$ [19–22].

Z-score can be used to calculate the normalization.

$$Z - \text{score} = \frac{x - \mu}{\text{S. D.}}$$

where

$$x = \text{score}$$
$$\mu = \text{mean}$$
$$\text{S. D.} = \text{standard deviation}$$

2. **Confidence Interval**: It is a type of interval estimate from the sampling distribution that gives a range of values where the population statistic may lie.

$$C.I = \mu \mp Z * \frac{\sigma}{\sqrt{n}}$$

Example 1: In Figure 3.4, we can observe the bell-shaped symmetrical curve with the mean in between the confidence interval. It is an example of the schematic diagram of how data is distributed. The confidence interval is 95%. Now, in this example, $Z_{95\%}$, is given as 1.96. To find the confidence interval, we will apply the formula of C.I. given above. Suppose the given $\mu = 40$, $\sigma = 40$, $n = 100$.

Note: The confidence Interval can be of any value. Here, we have taken it as 95% because the probability that the Confidence Interval will contain the true population mean is likely high.

Applying the formula for C.I.:

$$C.I = \left(40 - 1.96 * \frac{40}{\sqrt{100}}\right)\left(40 + 1.96 * \frac{40}{\sqrt{100}}\right)$$

$$C.I = (32.16)\,(47.84)\,(\text{Answer})$$

3. **Margin of Error**: It is half of the confidence interval. It can be defined as the sampling error by a person who collected the data. If the sample mean lies in the margin of error, then it is possible that its actual value is equal to the population means and the difference occurred by chance. Anything outside the margin of error is considered statistically significant. The margin of error is on either side of the mean; it can be both positive and negative. The greater the sample size, the smaller the confidence interval — it leads to a more accurate population mean from the sample means, and vice versa. There are different confidence intervals for different sample means. By 95% confidence interval, it does not mean that the probability of a population mean will lie in an interval of 95%; instead, it means that 95% of interval estimates will contain the population statistic [23,24].

4. **Probability Density Function**: A function used to represent the probability distribution of a continuous random function.

Important points:

   i. A normal curve is symmetrical, unimodal, and bell-shaped.
  ii. In a standard normal distribution, the mode is the same as the mean and the mean is 0 for a standard normal distribution.
 iii. The area to the left of ($\mu + \sigma$) of a normal distribution is approximately equal to 0.84.

5. **Hypothesis**: A hypothesis is nothing but a proposed explanation for a phenomenon.
   The null hypothesis means the sample statistic must be equal to the population statistic or that the intervention doesn't bring any significant difference to the sample. An alternate hypothesis negates the null hypothesis; it concludes that the intervention brings a significant difference to the sample, or that the sample is significantly different from the population. Hypothesis testing is done on various levels of confidence and makes use of a z-score to calculate the probability. One cannot accept the null hypothesis — only reject it or fail to reject it. As a practical tip, the null hypothesis is framed on the statement one wants to disprove. A person might make errors while making conclusions using hypothesis testing [25,26].
6. **Critical Value**: A critical value is a point on the scale of the test statistic beyond where one rejects the null hypothesis and is derived from the level of significance of the test.
7. **Directional Hypothesis**: In this hypothesis, the null hypothesis is tested only in one direction (as shown in Figure 3.5).
8. **Non-Directional Hypothesis**: In this hypothesis, the null hypothesis is tested in both directions (as shown in Figure 3.6).
9. **T-Tests**: These tests which are similar to z-score tests use standard deviation to estimate population standard deviation. These tests include one-tailed and two-tailed tests. To perform a test, we will first have to define the null and alternate hypotheses. Then, computation of the t-statistic will begin, the t-

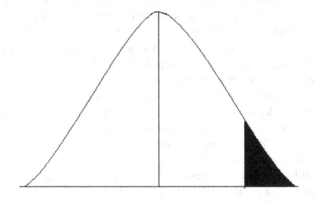

FIGURE 3.5 Directional Hypothesis.

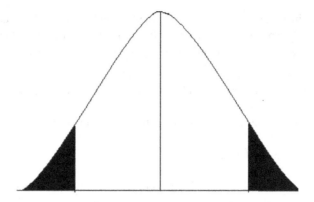

FIGURE 3.6 Non-Directional Hypothesis.

critical value from the t-table shall be obtained, and if the t-statistic value is less than the t-critical value, then there is no difference in the sample population and population distribution [27].

$$T - \text{statistic} = \frac{x(bar) - u}{s/\sqrt{n}}$$

10. **Correlation:** It is used to determine the relationship between two variables. It is an important topic of machine learning and statistical analysis. It is denoted by r. The value ranges from -1 to $+1$. Here, 0 means no correlation.

5 INTRODUCTION TO PREDICTIVE MODELLING

Predictive Modelling is making use of past data and other attributes to predict the future [28].

It lets us know the difference between Supervised and Unsupervised Learning before diving deep into types of predictive models.

Supervised learning is where we have both the input data set and output data set, as well as a method of mapping input to that of output — it is complex in nature.

The number of classes is known, and supervised learning is more predictable because there is a correct answer, and the algorithm stops as soon as it reaches the result successfully.

Unsupervised learning acts as if there is no correct answer. It has less computational complexity and the number of classes is not known. Unsupervised learning is where we only have the input data sets and not the output data set.

Two types of predictive models:

I. Supervised Learning: It contains two phenomena.
   a. Regression – Regression is a statistical method used in finance, investing, and other disciplines that attempt to determine the strength and character of the relationship between one dependent variable (usually denoted by Y) and a series of other variables (known as independent variables).
   b. Classification – Classification is a process of categorizing a given set of

data into classes. It can be performed on both structured and unstructured data. The process starts with predicting the class of given data points. The classes are often referred to as target, label, or categories.

II. Unsupervised Learning: It also contains two phenomena.
   a. Document clustering – This phenomenon is used to cluster together similar documents.
   b. Market segmentation – Segmentation of target people into groups.

Stages of Predictive Modeling:

i. **Problem Definition**: Identify the problem statement and accordingly formulate the steps to solve them.
ii. **Hypothesis Generation**: After listing down all possible variables that may influence the problem objectively, consider generating a hypothesis. Hypothesis generation is done before looking at the data; it helps us list down all the factors which might affect the problem without being biased. Prevents waste time analyzing all available data.
iii. **Data Extraction/Collection**: The process of extraction of various data from various sources and using them to manipulate for better data visualization.
iv. **Data Exploration and Transformation**: The steps are:

A. Reading the raw data: There are two common types of file formats
   a. CSV Format (.csv)
   b. Excel format (.xls/.xlsx)
   The steps are:

   a. Reading the CSV file- (read_csv())
   b. Size of data- (shape)
   c. Name of the variables in the data set- (columns)

Program 1.13
```
df=pd.read_csv()
df.shape
df.columns
df.head()
```

   d. Look at the top five observations in the dataset- (head())
      To know the required dimensions of the data set, it is necessary to remember some small snippets of codes. Refer to Program 1.13

B. Variable Identification:

Process of identifying which variables are:

a. Independent and dependent variables: Dependent variables are what one is trying to predict. Independent variables are those variables that help predict the dependent variable.
b. Continuous and categorical variables: Variables that are discrete in nature are known as categorical and those having an infinite number of possible values are known as continuous. Categorical variables are stored as objects and continuous variables are stored as int/float.

Reasons for variable identification:

I. Techniques like supervised learning require identification of the dependent variable.
II. Different data processing techniques for categorical and continuous data.

C. Univariate analysis: Analysis of variables one by one.

The main points are:

a. Exploring one variable at a time.
b. Summarize the variables.
c. Make sense out of the summary to discover insights, anomalies, etc.

For continuous variable:

a. Central tendency and dispersion
b. Distribution of variable
c. Presence of missing values
d. Presence of outliers

For categorical variable:

a. Count – Absolute frequency of each category in a categorical variable.
b. Count % – Proportion of different categories in a categorical variable expressed in %.

D. Bivariate analysis: Analysis of the relationship between two variables. This includes the phenomenon when two variables are studied together for their empirical relationship.

Reasons for Bivariate Analysis:

a. Helps in prediction (when two variables are associated, one may be used to infer the other).
b. It helps detect anomalies.

Types of bivariate:

a. Continuous – continuous variable
   Example: Age-fare price
b. Categorical – continuous variable
   Example: Gender-age price
c. Categorical – categorical variable

Example Gender-survived

E. Missing Value Treatment: Missing values refer to those absent in the data set. Sometimes, it becomes difficult to train and test a set of data that has missing values [29,30]. There may be many reasons for it. Some of them are:
   a. Nonresponse
   b. Error in data collection
   c. Error in reading data

There are three types of missing values:

a. Missing completely at random (MCAR)
b. Missing at random (MAR)
c. Missing not at random (MNAR)

Missing completely at Random (MCAR): These missing values have no relation with the variable where missing values exist and other variables in a dataset.

Missing at Random (MAR): These missing values have no relation to the variable in which missing values exist, but they have a relation with other variables.

Missing not at Random (MNAR): These missing values have a relation to the variable in which they are missing.

There are some of the functions in Python that are used to identify the several types of variables such as:

a. describe(): It can only be used to identify values in the continuous variable.
b. Isnull(): It is used in both categorical and continuous variables.

After identifying the missing values, one needs to deal with them properly. There are two methods to do this — imputation and deletion. Imputation contains two processes, each for continuous and categorical variables. For the categorical variables, mode and

Program 1.14
```
file=pd.read_CSV("data.csv")
file.shape
file.describe() # count helps in detecting missing values
file.isnull() # missing values represented by True file.isnull.sum() #gives the total no. of missing values.
file.dropna() # drops all the rows which contain missing values.
```

classification model are used; for continuous variable, mean, median, and regression model are used. Deletion deals with two types: row-wise deletion and column-wise deletion. To perform the operations for the treatment of missing values, various small snippets of code must be remembered. Refer to Program 1.14".

Dropping rows which contain missing values is also an important task to provide as a data scientist. Refer to Program 1.15.

Program 1.15
```
file.dropna() # drops all the rows which contain missing values.
file=file.dropna() #affects original file.
file.dropna(how='all') #drop the rows where all the entries are missing.
file.dropna(axis=1) #deletes the column which contain any no. of missing values.
```

Wherever the missing values exist is yet another essential task to be implemented, The Python platform provides features to help deal with it. In the given Program 1.16, we assume that the file contains a column named 'Age'. So, we fill the missing values with certain other values in the column 'Age'. One can choose any other column if they want.

F. Outliers Treatment: Any value that falls outside the range of data is termed as an outlier. They differ significantly from the other observations. It may be present in a data set due to reasons such as:
   a. Data entry errors
   b. Measurement errors
   c. Processing errors.

Program 1.16
```
file.fillna(0) #all missing values are filled with 0.
file['Age'].fillna(0) #fill zero only in age column
file['Age'].fillna(file['Age'].mean()) #fill the missing values with mean of age.
```

There are two types of Outliers:

   a. Univariate Outliers – if analyzing one variable
   b. Bivariate Outliers – if analyzing two variables

There are various kinds of methods where a person can identify an outlier. Outliers are extremely easy to identify as they stand out from the rest of the data; they always differ in the range of their existing values. The two most common methods of identifying outliers are the Box plot and the Scatter Plot. Another manner of identification is the use of The Formula method that uses the range of the values in the data set. The formula is: <Q1 – 1.5 * IQR or >Q3 + 1.5 * IQR, where IQR is the Interquartile Range [31]. If any values fall in this range, then they are considered outliers. Treating outliers is another mammoth task to do. It is, though, like dealing with missing values. Some of the ways are:

a. Deleting observations
b. Transferring and binning values
c. Input outliers like missing values
d. Treat them separately

To perform the treatment of outliers, refer to Program 1.17.

Program 1.17
```
df['age'].plot.box() #creates box plot
df.plot.scatter('age','fare') #creates scatter plot
df[df['fare']<300] #First treatment- removing outliers
df.loc[df['age']>65, 'age']=np.mean(df['age']) #re-
placing outliers with mean of age
```

G. Variable Transformation: Process by which we replace a variable (may be x-axis variable or y-axis variable) with another function of that variable (e.g., replacing x with its logarithm).

Reasons for using variable transformation:

a. To change the scale of variable.
b. To transform a non-linear relationship to a linear relationship.
c. Create symmetric distribution from skewed distribution.

Some common methods of performing variable transformation:

a. Logarithm: taking the log of a variable reduces the right skewness of the variable.
b. Square Root: used for the right-skewed variable only with the values.
c. Cube Root: used for right-skewed with positive or negative values.
d. Binning: used to convert continuous variables into a categorical variable.

Variable Transformation is used to adjust the schematic diagrams visually. If a histogram of a certain kind of data is too right-skewed or left-skewed and one needs to readjust it so that it becomes visually appealing, one may utilize variable transformation. It is also used to change the relationship between the x-axis and y-axis variables. To change the relation to either logarithmic or square root (whichever more compelling), variable transformation can be used. Use Program 1.18 to notice the same [32].

Statistical Methods

Program 1.18
```
df=pd.read_csv('data.csv')
df['Age']plot.hist()   #plots  histogram(right-skewed histogram)
np.sqrt(df['age']).plot.hist  #plots histogram which fits the range.
```

v. **Predictive Modeling**: Data science is a blend of both statistical analysis and machine learning. This field deals with the manipulation and eventual utilization of data. It shows that the data collected from the past can be helpful in judging or predicting future consequences. In this regard, we divide the data set into two parts:
   a. Training Data
   b. Testing Data

The training data is the data used to train a model that will help in predicting future results. The model feeds in the training data to train itself and applies it over the test data. The test data is used to check the accuracy of the model. Therefore, when we want to check the error rate of the model, we need the actual answer so that we can compare it with the result. There is also another method to check for error. If the training data score matches or is somewhere equal to the test data score, then there is minimal error.

In the following Python code, we will show how the data are trained, tested, and checked for error. For the programs, import pandas matplot library [33].

In Program 1.19, we are setting aside the first 8,000 data entries to train the model and the rest to test the model. Now, we need to remove the dependent variables from the data set so the model can only be made from independent variables.

Program 1.19
```
data=pd.read_csv("train.csv")
train=data[0:7999] #training
test=data[8000;] #testing
```

In Program 1.20, we are setting aside the dependent variable (i.e. 'Item_Outlet_Sales'). Now, we will assign it to a different variable (i.e. y_train).

Program 1.20
```
x_train=train.drop('Item_Outlet_Sales',axis=1)
```

Program 1.21
```
y_train=train['Item_Outlet_Sales',,axis=1) #contains
dependent variable
```

Like the training data set, the test data set also needs to have only independent variables. So, the same shall be done.

Program 1.22
```
x_test=test_drop('Item_Outlet_Sales',axis=1)
true_p=test('Item_Outlet_Sales')
```

Then, as the library sklearn is being used [34], it only accepts input in the form of numbers and not string. So, one must be careful of any string data type values in the data set. First, the linear regression model is made by creating an object. (Program 1.23)

Program 1.23
```
lreg=LinearRegression() #creates object
lreg.fit(x_train,y_train) #tries to fit the data
```

Then, the dummies are created to deal with it. (Program 1.24)

Program 1.24
```
x_train=pd.get_dummies(x_train) #creation of dummies
x_train.shape(7999,1604)
x.test=pd.get_dummies(x_test)
```

As previously explained, treatment of missing values must be done always. For this, refer to Program 1.25

Program 1.25
```
x_train.fillna(0,inplace=true)     #treating missing
values
x_test.fillna(0,inplace=true)
lreg.fit(x_train,y_train) #success
lreg.predict(x_test)
```

Finally, we shall predict the data and check for accuracy. Refer to Program 1.26

Program 1.26
```
pred=lreg.predict(x_test) #success
lreg_score(x_test,true_p) #0.4044
lreg_score(x_train,y_train) #0.6489
```

By running the values for a sample data, the test data score is almost similar to the training data score, yet some error exists. The score for the test data is 0.4044, whereas the training data score comes around 0.6489.

6 CONCLUSION

It is very handy to use these Python libraries and functions to statistically make, predict, and check a model that may open various paths towards the applications of machine learning. These snippets of codes will surely help achieve the idea of making a predictive model. This chapter extensively combines the importance of statistical analysis along with the methods of modeling. Various definitions have been provided to familiarize the reader with data-related terminologies. This chapter has successfully tried to make a clear understanding of how data and its manipulation works. This chapter also used several references, terms, and concepts related to machine learning that one may find relevant. Any new implementation of the concepts above can help create a structured model that could help fields such as medicine, agriculture, research, and software development.

REFERENCES

1. Delescluse, M., Franconville, R., Joucla, S., Lieury, T., and Pouzat, C. (2012). Making neurophysiological data analysis reproducible: Why and how? *Journal of Physiology-Paris, 106*(3–4), pp. 159–170.
2. McMurdie, P. J. and Holmes, S. (2013). Phyloseq: An R package for reproducible interactive analysis and graphics of microbiome census data. *PloS One, 8*(4).
3. Shi, L., Perkins, R. G., Fang, H., and Tong, W. (2008). Reproducible and reliable microarray results through quality control: good laboratory proficiency and appropriate data analysis practices are essential. *Current Opinion in Biotechnology, 19*(1), pp. 10–18.
4. Baumer, B. (2015). A data science course for undergraduates: Thinking with data. *The American Statistician, 69*(4), pp. 334–342.
5. McShane, L. M., Radmacher, M. D., Freidlin, B., Yu, R., Li, M. C., and Simon, R. (2002). Methods for assessing reproducibility of clustering patterns observed in analyses of microarray data. *Bioinformatics, 18*(11), pp. 1462–1469.
6. Miller, R. B. (1994). *Statistics for business: Data analysis and modeling.* Duxbury Press.
7. Raschka, S. (2015). *Python machine learning.* Packt Publishing Ltd.

8. Härdle, W., Lu, H. H. S., and Shen, X. (Eds.). (2018). *Handbook of big data analytics.* Springer International Publishing.
9. Oja, H. (1983). Descriptive statistics for multivariate distributions. *Statistics & Probability Letters, 1*(6), pp. 327–332.
10. Bruce, P. and Bruce, A. (2017). *Practical statistics for data scientists: 50 essential concepts.* O'Reilly Media, Inc.
11. Lutz, M. (2001). *Programming python.* O'Reilly Media, Inc.
12. Cleveland, W. S. (2001). Data science: an action plan for expanding the technical areas of the field of statistics. *International Statistical Review, 69*(1), pp. 21–26.
13. Reid, N. (2018). Statistical science in the world of big data. *Statistics and Probability Letters, 136*, pp. 42–45.
14. Nelder, J. A. (1999). From statistics to statistical science. *Journal of the Royal Statistical Society: Series D (The Statistician), 48*(2), pp. 257–269.
15. Ahn, S. and Fessler, J. A. (2003). *Standard errors of mean, variance, and standard deviation estimators* (pp. 1–2). EECS Department, The University of Michigan.
16. VanderPlas, J. (2016). *Python data science handbook: Essential tools for working with data.* O'Reilly Media, Inc.
17. Punch, W. F., & Enbody, R. (2012). *Practice of computing using Python.* The Addison-Wesley Publishing Company.
18. Lowry, R. (2014). Concepts and applications of inferential statistics. 2003. URL: http://vassarstats. net/textbook/ch14pt2. html.
19. Panda, S. K. and Jana, P. K. (2018). Normalization-based task scheduling algorithms for heterogeneous multi-cloud environment. *Information Systems Frontiers, 20*(2), pp. 373–399.
20. Mishra, S. K., Puthal, D., Sahoo, B., Jena, S. K., and Obaidat, M. S. (2018). An adaptive task allocation technique for green cloud computing. *The Journal of Supercomputing, Springer, 74*(1), pp. 370–385.
21. Panda, S. K., Gupta, I., and Jana, P. K. (2019). Task scheduling algorithms for multi-cloud systems: Allocation-aware approach. *Information Systems Frontiers, 21*(2), pp. 241–259.
22. Mishra, S. K., Puthal, D., Sahoo, B., Sharma, S., Xue, Z., and Zomaya, A. Y. (2018). Energy-efficient deployment of edge data centers for mobile clouds in sustainable IoT. *IEEE Access, 6*, pp. 56587–56597.
23. Anderson, D. R., Burnham, K. P., and Thompson, W. L. (2000). Null hypothesis testing: problems, prevalence, and an alternative. *The Journal of Wildlife Management*, pp. 912–923.
24. Mishra, S. K., Sahoo, S., Sahoo, B., and Jena, S. K. (2020). Energy-efficient service allocation techniques in cloud: A survey. *IETE Technical Review, 37*(4), 339–352.
25. Müller, A. C. and Guido, S. (2016). *Introduction to machine learning with Python: a guide for data scientists.* O'Reilly Media, Inc.
26. Van Der Aalst, W. (2016). Data science in action. In *Process mining* (pp. 3–23). Springer, Berlin, Heidelberg.
27. Demšar, J. (2008). On the appropriateness of statistical tests in machine learning. In Workshop on Evaluation Methods for Machine Learning in conjunction with ICML (p. 65).
28. Alpaydin, E. (2020). *Introduction to machine learning.* MIT Press.
29. McKinney, W. (2015). *Pandas, python data analysis library.* see http://pandas.pydata. org.
30. McKinney, W. and Team, P. D. (2015). *Pandas—powerful Python data analysis toolkit* (p. 1625).

31. Diggle, P. J. (2015). Statistics: a data science for the 21st century. *Journal of the Royal Statistical Society: Series A (Statistics in Society)*, *178*(4), pp. 793–813.
32. Gentleman, R. and Temple Lang, D. (2007). Statistical analyses and reproducible research. *Journal of Computational and Graphical Statistics*, *16*(1), pp. 1–23.
33. Tosi, S. (2009). *Matplotlib for Python developers*. Packt Publishing Ltd.
34. Pedregosa, F., Varoquaux, G., Gramfort, A., Michel, V., Thirion, B., Grisel, O., ... Vanderplas, J. (2011). Scikit-learn: Machine learning in Python. *Journal of Machine Learning Research*, *12*(Oct), pp. 2825–2830.

4 An Approach for Energy-Efficient Task Scheduling in Cloud Environment

Mohapatra Subasish[1], Hota Arunima[2], Mohanty Subhadarshini[3], and Dash Jijnasee[4]

[1]Department of Computer Science and Engineering, College of Engineering and Technology, Bhubaneswar, India, Email: smohapatra@cet.edu.in

[2]Department of Computer Science and Engineering, College of Engineering and Technology, Bhubaneswar, India, Email: arunimahota123@gmail.com

[3]Department of Computer Science and Engineering, College of Engineering and Technology, Bhubaneswar, India, Email: sdmohantycse@cet.edu.in

[4]Department of Computer Science and Engineering, College of Engineering and Technology, Bhubaneswar, India, Email: jijnasee.nit08@gmail.com

1 INTRODUCTION

In the twenty-first century, the advancement of innovation has brought about significant usage of distributed computing as a pay-per-use model. This is because of its highlight features like multi-tenancy, scalability, agility, mobility, and resource utilization using the virtualization procedure. In this pay-per-use model, end-users do not require the purchase of any software to perform a task—the only requirement is a paid internet connection for the duration of its use. This type of facility reduces cost and encourages the use of cloud resources dynamically. Cloud provides diverse types of services depending on the demand of end-users [1]. Among all the advantages that the cloud offers, there still arise some complexities and obstacles with the supply of virtual machines—these must be removed by the cloud service provider. In cloud facilities, we must consider response time, execution time, makespan, power consumption, effective resource utilization, cost, and—most importantly—load balance. Load balancing is a technique to distribute workloads among the servers, networks, etc. virtually [2]. It aims to minimize response time, cost; maximize throughput, and resource utilization by eliminating the overuse of resources, which enhances the reliability of the system. It follows two types of approaches: static and dynamic load balancing. In static load balancing, it does not check the current status of the task—the tasks are not pre-empted.

Hence, this approach is not used widely. In dynamic load balancing, the tasks always come in a pre-empted order, so it is easy to regulate which systems are underutilized, over-utilized, or remain idle. The main problem in dynamic load balancing is the effective utilization of resources, energy consumption, and performance of the system. Most approaches would focus on the equal distribution of load across multiple servers, which ultimately increases response time rather than energy consumption in the system. Data centers are the main part of the cloud that consists of many servers. The virtualization technique is implemented on these physical servers to make them function in a virtualized manner. Type 2 hypervisors are used on these from where many virtual machines are made according to the use of the system. These virtual machines and their corresponding servers and data centers consume more energy, which leads to an increase in cost and carbon dioxide emission. The government put emphasis on the use of green technology by reducing the higher consumption of energy. Considering the requirements, the authors have proposed an enhanced load balancing algorithm that minimizes response time, maximizes resource utilization, and minimizes energy consumption.

2 LITERATURE SURVEY

Load balancing is a technique to allocate the task equally among servers, minimizing response time, and maximizing the throughput so that VM allocation will be performed successfully. It has also provided benefits by reducing energy consumption and cost, which push this technology towards the green computing era. This process also helps increase efficiency and reduce the workload. Many researchers have worked on this process by using different approaches such as heuristic, metaheuristic, and hybrid. A modified PSO and improved Q-learning algorithm have been proposed to improve throughput, minimize the makespan, and efficiently utilize the energy [3]. The performance of the algorithm is illustrated by how independent tasks are migrated, the delayed and response time of the task, and the result of the makespan before and after the load balance. The simulations are carried out using CloudSim. Fuzzy controlled approaches have been proposed to efficiently balance the load and reduce the cost and energy of data centers [4]. The performance of the algorithm is stated as an offline geographical load balancing approach to map the task nonlinearly and send it to the data center [5]. An online power management and scheduling algorithm has been proposed to optimize the cost and power of data centers. It put focus on the Lyapunov optimization theory to manage the carbon footprint effortlessly in data centers. A novel energy-efficient algorithm has been developed to optimize the energy consumption in servers [6]. Here, the algorithm distributes tasks parallel to their respective resources in the cloud. The experiment result is carried out using CloudSim. A heuristic clustering-based load balancing algorithm has been proposed to efficiently balance the load [7]. This approach is based on the Bayes algorithm for better performance. A genetic algorithm-based load balancing algorithm has been proposed to reduce the energy consumption in data centers [8]. Here, the algorithm is compared with the max-min algorithm and it is concluded that the genetic algorithm gives better results as compared to the max-min when tasks are assigned to virtual machines to reduce energy consumption. A dynamic load balancing algorithm has been proposed to optimize response time, storage, and reduced energy consumption [9]. Here, an improved weighted round-robin algorithm that works at the completion time of each task is proposed; after that, it equally distributes the task to vms. This algorithm worked homogeneous as well as heterogeneous tasks and vms. A dynamic load balancing policy-based algorithm has been implemented to reduce the power consumption and utilize the resources effectively in the cloud [10]. The experiment is carried out in CloudSim. A survey is carried out by taking all the approaches of load balancing to maximize the resource utilization, optimize response time, makespan, and reduce power consumption [11]. A dynamic load balancing approach has been proposed to reduce makespan and utilize resources effectively [12]. This algorithm is based on the task migration approach where the energy consumption of tasks is calculated first before it goes to vm, then a comparison is carried out by taking a different heuristic approach-based algorithm. A job scheduling algorithm has been proposed in the cloud which focuses on the priority and cost of the job [13]. A survey is carried out on load balancing approaches to provide a quick review of algorithms and their outcomes [14]. Here, the heuristic approaches are performed on a CloudSim simulator. Apart from this fuzzy logic, Hadoop-based algorithms are proposed to minimize makespan and energy consumption in data centers, maximizing throughput and resource allocations. The main concern in load balancing approaches is to reduce response time and power, efficiently allocating the resource; some researchers have approached a dynamic mechanism for this [15]. Here, the dynamic mechanism has worked on several tasks that come from the client site and it was observed that increasing tasks would not affect the response time at both heterogeneous and homogeneous data centers. An energy-efficient strategy has been proposed to allocate the vm in the cloud [16]. In this strategy, the authors have taken the hybrid approach as well as a best-fit algorithm to effectively allocate each task to respective virtual machines so that they are never in an idle state. A multi-objective framework has been developed to reduce the energy and provide security in the cloud [17]. This framework provides better security to the servers and achieves better results in resource allocation. A heuristic approach-based algorithm has been proposed for the IaaS cloud [18]. They have also worked on the resource allocation in the host and proper scheduling on virtual machines. The experiment is carried out in CloudSim. A multidimensional load balancing approach has been proposed in the cloud to improve resource scheduling efficiently [19]. The approach is focused on removing underutilized and over-utilized resources and improving the latency of the task. A taxonomy model has been developed to allocate the resources and balance the load across data centers [20]. Here, it put the focus on different metrics of energy consumption. An effective task scheduling approach that pushes the cloud in the green technology era has been proposed [21]. The approach puts emphasis on the overuse of power consumption and performance criteria. A clonal selection theory is applied to optimize power consumption. A survey is carried out on different approaches and the latest technology to effectively optimize energy consumption and the cost of data centers [22] (Figure 4.1).

3 PROPOSED MODEL

The client sends a request for service to the cloud. Requests come in the form of tasks entered into the cloud service provider, where it decides whether it will be capable to perform its operation. If it is, then the service provider will send the request to the cloud

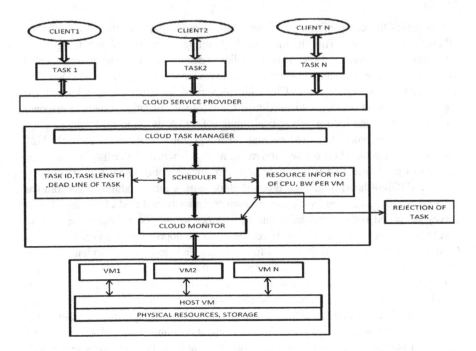

FIGURE 4.1 Proposed Model for Load Balancing.

task manager. Inside the cloud task manager, tasks are arranged according to length and deadline. After that, it will enter into a scheduler where a scheduling algorithm is present to plan the task while adding some resource information such as cpu and bandwidth per virtual machine. If for any reason tasks are performing its operation beyond deadline, then the scheduler rejects that task and continues its operation. The scheduler connects with physical resources through the help of a virtual cloud monitor.

4 PROBLEM FORMULATION

We schedule the tasks with the virtual machine in a way that cloud user can take minimum time to execute their task and use the average resource utilization. Cloud task scheduler receives t number of task requests T1, T2, T3, ..., Tn. All the tasks are independent in nature and tasks with deadlines that have passed are automatically rejected by the scheduler. When one task is allocated to the virtual machine, it executes the complete task before it executes the next task. Every task has a length, which is expressed in MI (million instructions); each task requires p processing speed, q number of cpu, r amount of main memory, and required bandwidth $B_{j,k}$ between the virtual machine j and k in MBPS. Cloud task scheduler contains the information about the m heterogeneous virtual machine VM1, VM2, VM3, ..., VMm, that have different processing power in terms of processing speed, memory, bandwidth, hard disk, etc. First, we calculate the capacity of individual virtual machine and capacity of all virtual machine using the formula given in [23].

An Approach for Energy-Efficient

Capacity of virtual machine can be achieved by with the help of (4.1)

$$CVM = \text{processing speed} \times \text{number of cpu} + \text{bandwidth of vm} \qquad (4.1)$$

Load information at the virtual machine can be achieved by load of virtual machine:

$$LVM = \frac{\text{No. of task} \times \text{task length}}{\text{Processing speed} \times \text{no. of cpu}} \qquad (4.2)$$

Execution time of a task at virtual machine can be found with the help of (4.3):

$$ET = \frac{\text{Task length}}{\text{Processing speed} \times \text{no. of cpu}} \qquad (4.3)$$

Makespan of the virtual machine can be achieved by

$$MVM = \text{Max}\left(\text{Min} \sum_{i=1}^{m} ET\right) \qquad (4.4)$$

Since the total task completion time can be expressed as the sum of current load of $M_i(CL_i)$ and a new load of $M_i(NL_i)$, that is:

$$T_i = CL_i + NL_i \qquad (4.5)$$

Utilization can be achieved by dividing the task computation time of each VM by the makespan value. The utilization of the individual VM (UMi) can be given by:

$$UM_i = \frac{T_i}{MVM} \qquad (4.6)$$

where T is the task completion time. Similarly, avg. utilization can be calculated as:

$$AUM_i = \frac{\sum_{i=1}^{m} UM_i}{m} \qquad (4.7)$$

To facilitate the design of dynamic algorithm for load balancing, the above three equations are incorporated into a single fitness function:

$$\text{Fitness} = \frac{1}{\text{Makespan}} \times AUM_i \qquad (4.8)$$

The modify operation is carried out after the fitness value is assigned to everyone in the population [24]. Modify equation:

$$X'_{j,k,i} = X_{j,k,i} + r_{1,j,i}\left[X_{j,best,i} - |X_{j,k,i}|\right] - r_{2,j,i}\left[X_{j,worst,i} - |X_{j,k,i}|\right] \quad (4.9)$$

where $X'_{j,k,i}$ is the updated value of $X_{j,k,i}$. Here is the value of jth variable for the kth candidate during ith iteration. $r_{1,j,i}$ and $r_{2,j,i}$ are the two random numbers for the jth variable during ith iteration in the range [0,1]. The term $r_{1,j,i}(X_{j,best,i} - |X_{j,k,i}|)$ indicates the tendency of the solution to move closer to the best solution. $-r_{2,j,i}(X_{j,worst,i} - |X_{j,k,i}|)$ indicates the tendency of the solution to avoid the worst solution. $X'_{j,k,i}$ is accepted if it gives better function value than $X_{j,k,i}$. All the accepted function values at the end of iteration are maintained and become the input to the next iteration. The process is continued until all the particles achieve the global optimized value.

5 PROPOSED ALGORITHM

We proposed an energy efficient task scheduling algorithm in which our main aim is to reduce makespan and energy consumption and enhance resource utilization in cloud. Here, we took t number of tasks. Length of t number of tasks is generated randomly from 20,000MI to 500,000MI. Also assign deadline to each task. After that, we created v number of virtual machines in terms of speed and memory capacity. Both the tasks and virtual machines are sorted in decreasing order according to their speed and length using bubble sort algorithm. Once the tasks and virtual machines are sorted, the allocation of task to virtual machine is performed by using dynamic Jaya algorithm as mentioned in step 3. If any task exceeds its deadline during allocation, then that task will be rejected. After the allocation load balancing operation has been carried out, check the status of virtual machines. We illustrated the status using underload and overload. If the virtual machine is less than 20% of its capacity, then goes to the underloaded type and if the virtual machine is greater than 80% of its capacity, then it will go to the overloaded type. We sorted the overloaded virtual machine in decreasing order and under loaded virtual machine in increasing order. After that, we transferred the task from overloaded virtual machine to underloaded virtual machine and calculated the transfer time of each task. We also checked the status of the virtual machine until the load balancing operation has been performed successfully.

1. Generate task ID, length, and deadline. Arrange the tasks in decreasing order.
2. Generate virtual machines and arrange them in decreasing order.
3. Allocate the task to virtual machine using (4.5) to (4.9).
4. If the task cannot complete its execution in time, then it will be rejected.
5. Find the load of virtual machine using (4.1) and (4.2).
6. Check the status of virtual machine (i.e. underloaded virtual machine = 0.2 * CVM; overloaded virtual machine = 0.8* CVM)
7. Sort the underloaded virtual machine in increasing order and overloaded virtual machine in decreasing order.
8. While overloaded virtual machine! =null && under loaded virtual machine!=null transfer task From overloaded to under loaded virtual machine.

TABLE 4.1
Characteristics of the Virtual Machine

VMId	VM MIPS	VM Image Size	Memory	No. of CPU	VMM
0	580	1000	256	1	XEN
1	1000	1000	512	1	XEN
2	500	1000	256	1	XEN
3	700	1000	512	1	XEN
4	400	1000	256	1	XEN
5	250	1000	256	1	XEN

9. Calculate transfer time of task (i.e. $\frac{\text{Task length}}{\text{Bandwidth}}$).
10. After task, transfer check the status of each virtual machine. If any virtual machine is still in overloaded condition, repeat load balancing operation again.

6 SIMULATION AND EXPERIMENTAL RESULT

Makespan value is indirectly proportional to average resource utilization, so we proposed and implemented our algorithm in CloudSim simulator to optimize the makespan and maximize the utility of resources by taking six virtual machines shown in Figures 4.1 and 4.2. We ran the simulation for more than 200 times on different tasks (i.e. range from 10 to 50). Table 4.1 shows the virtual machine properties (i.e., we took six heterogeneous virtual machines for the simulation experiment); all the virtual machines have different processing power and memory. Table 4.2 shows task properties where we generated ten tasks of random length with specified range. All tasks are independent from each other.

TABLE 4.2
Characteristics of the Task

Cloudlet ID	Length	No. of CPU
0	75424	1
1	25572	1
2	240802	1
3	395311	1
4	261266	1
5	75433	1
6	209110	1
7	252925	1
8	39660	1
9	20711	1

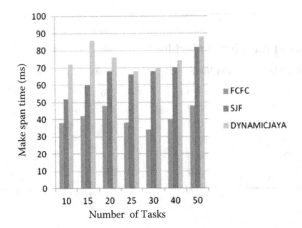

FIGURE 4.2 Makespan Time Calculation.

We calculated the experiment with six virtual machines and ten tasks at the start, then we increased the task to 10-15, 20, and so on at random length. We calculated the makespan time as well as resource utilization ratio and the makespan time of virtual machines, and compared the results with other existing algorithms. We have taken two results to prove that proposed algorithm is better than other existing algorithms. Results shown in Figures 4.2 and 4.3 prove that proposed algorithm perform better than existing algorithms.

Table 4.3 shows makespan comparison proposed algorithm with existing algorithm at virtual machine 6 (Table 4.4).

Figure 4.2 illustrates the comparison of two existing algorithm with the proposed algorithm based on makespan.

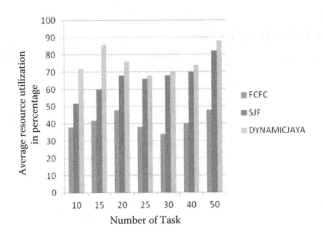

FIGURE 4.3 Average Resource Utilization Ratio.

TABLE 4.3
Comparison of Makespan of Different Algorithms

FCFC	SJF	Dynamic Jaya
680	563	520
660	714	549
785	730	575
880	750	589
840	790	615
1399	1127	915
1541	1326	1083

TABLE 4.4
Average Resource Utilization Data

FCFS	SJF	Dynamic Jaya
38	52	72
42	60	86
48	68	76
38	66	68
34	68	70
40	70	74
48	82	88

Table 4.4 shows the comparison of average resource utilization between the proposed algorithm and existing algorithm at virtual machine 6.

Figure 4.3 illustrates the comparison of average resource utilization with the existing algorithm and the proposed algorithm.

7 CONCLUSION AND FUTURE WORK

In this paper, a dynamic green optimized energy scheduling using dynamic load balancing technique has been discussed. Here, VMs are selected according to the SLA level given by the user (i.e. deadline). Also, to achieve high resource utilization and minimize the makespan, we compared our proposed algorithm with existing algorithms. It shows that our proposed algorithm works better than the existing. In the future, we aim to implement proposed algorithm with some hybrid approach to produce practical results and apply QoS requirements such as priority, privacy, security, etc.

REFERENCES

1. Elzeki, O. M., Reshad, M. Z., and Elsoud, M. A. (2012). Improved maxmin algorithm in cloud computing. *International Journal of Computer Applications Technology*, 50(12), pp. 22–27.
2. Lin, W., Wang, J. Z., Liang, C., and Qi, D. (2011). A threshold-based dynamic resource allocation scheme for cloud computing. *Procedia Engineering*, 23, pp. 695–703.
3. Jena, U. K., Das, P. K., and Kabat, M. R. (2020). Hybridization of meta-heuristic algorithm for load balancing in cloud computing environment. *Journal of King Saud University-Computer and Information Sciences*.
4. Toosi, A. N. and Buyya, R. (2015, December). A fuzzy logic-based controller for cost and energy efficient load balancing in geo-distributed data centers. In 2015 IEEE/ACM 8th International Conference on Utility and Cloud Computing (UCC) (pp. 186–194). IEEE.
5. Deng, X., Wu, D., Shen, J., and He, J. (2016). Eco-aware online power management and load scheduling for green cloud datacenters. *IEEE Systems Journal*, 10(1), pp. 78–87.
6. Moganarangan, N., Babukarthik, R. G., Bhuvaneswari, S., Saleem Basha, M. S., and Dhavachelvan, P. (2016). A novel algorithm for reducing energy-consumption in cloud computing environment: Web service computing approach. *Journal of King Saud University-Computer and Information Sciences* 28(1), pp. 55–67.
7. Zhao, J., Yang, K., Wei, X., Ding, Y., Hu, L., and Gaochao, X. (2015). A heuristic clustering-based task deployment approach for load balancing using Bayes theorem in cloud environment. *IEEE Transactions on Parallel and Distributed Systems*, 27(2), pp. 305–316.
8. Kar, I., Parida, R. N. R., and Das, H. (2016). Energy aware scheduling using genetic algorithm in cloud data centers. In 2016 International Conference on Electrical, Electronics, and Optimization Techniques (ICEEOT) (pp. 3545–3550). IEEE.
9. Parekh, M., Padia, N., and Kothari, A. (2016). Distance, energy and storage efficient dynamic load balancing algorithm in cloud computing. In 2016 3rd International Conference on Computing for Sustainable Global Development (INDIACom) (pp. 3471–3475). IEEE.
10. Acharya, S. and D'Mello, D. A. (2017). Energy and cost efficient dynamic load balancing mechanism for resource provisioning in cloud computing. *International Journal of Applied Engineering Research*, 12(24), pp. 15782–15790.
11. Thakur, A. and Goraya, M. S. (2017). A taxonomic survey on load balancing in Cloud. *Journal of Network and Computer Applications*, 98, 43–57.
12. Kumar, M. and Sharma, S. C. (2020). Dynamic load balancing algorithm to minimize the makespan time and utilize the resources effectively in cloud environment. *International Journal of Computers and Applications*, 42(1), pp. 108–117.
13. Kumar, M., Dubey, K., and Sharma, S. C. (2017). Job scheduling algorithm in cloud environment considering the priority and cost of job. In Proceedings of Sixth International Conference on Soft Computing for Problem Solving (pp. 313–320). Springer, Singapore.
14. Mishra, S. K., Sahoo, B., and Parida, P. P. (2020). Load balancing in cloud computing: A big picture. *Journal of King Saud University-Computer and Information Sciences*, 32(2), 149–158.
15. Acharya, S. and D'Mello D. A. (2017). Energy and cost efficient dynamic load balancing mechanism for resource provisioning in cloud computing. *International Journal of Applied Engineering Research*, 12(24), pp. 15782–15790.
16. Bharathi, P. D., Prakash, P., and Vamsee Krishna Kiran, M. (2017). Energy efficient strategy for task allocation and VM placement in cloud environment. In 2017 Innovations in Power and Advanced Computing Technologies (i-PACT) (pp. 1–6). IEEE.

17. Singh, A. K. and Kumar, J. (2019). Secure and energy aware load balancing framework for cloud data centre networks. *Electronics Letters*, *55*(9), pp. 540–541.
18. Adhikari, M. and Amgoth, T. (2018). Heuristic-based load-balancing algorithm for IaaS cloud. *Future Generation Computer Systems*, *81*, pp. 156–165.
19. Priya, V., Sathiya Kumar, C. and Kannan, R. (2019). Resource scheduling algorithm with load balancing for cloud service provisioning. *Applied Soft Computing*, *76*, pp. 416–424.
20. Kulshrestha, S. and Patel, S. (2019). A study on energy efficient resource allocation for cloud datacenter. In 2019 Twelfth International Conference on Contemporary Computing (IC3) (pp. 1–7). IEEE.
21. Lu, Y. and Sun, N. (2019). An effective task scheduling algorithm based on dynamic energy management and efficient resource utilization in green cloud computing environment. *Cluster Computing*, *22*(1), pp. 513–520.
22. Priyadarshini, R., Barik, R. K., and Mishra, B. K. (2020). Meta-heuristic and non-metaheuristic energy-efficient load balancing algorithms in cloud computing. In *Modern principles, practices, and algorithms for cloud security* (pp. 203–222). IGI Global.
23. Dhinesh Babu, L. D. and Krishna, P. V. (2013). Honey bee behavior inspired load balancing of tasks in cloud computing environments. *Applied Soft Computing*, *13*(5), pp. 2292–2303.
24. Rao, R. V. and Waghmare, G. G. (2017). A new optimization algorithm for solving complex constrained design optimization problems. *Engineering Optimization*, *49*(1), pp. 60–83.

5 Solar-Powered Cloud Data Center for Sustainable Green Computing

Saravanakumar A.[1] and Sudha M. R.[2]

[1]Center for Environmental Nuclear Research, Directorate of Research and Virtual Education, SRM Institute of Science and Technology, Kattankulathur-603 203, Chennai, Tamil Nadu, India
[2]Department of Computer Applications, Faculty of Science and Humanities, SRM Institute of Science and Technology, Kattankulathur 603-203, Chennai, Tamil Nadu, India

1 INTRODUCTION

Green cloud computing is changing the way we store and retrieve big data. It's an evolving technique in the world of data. Cloud computing has an ever-present paradigm in the world of technology—it allows us to do dayto-day tasks with ease as access to the information is excellent and easy to retrieve, store, and edit. Cloud computing has become known as an innovative technology and a virtualization model for the digital world. Mobile Cloud services provide a suitable on-demand service, broad network access, and resources with intended services in a highly customized manner with minimal supervision [1]. However, there is a growing concern to optimize the output of the cloud infrastructure and to make it more mobile to ease the burden on aging computer infrastructure. Computer infrastructure is becoming outdated and is causing major problems in the overall energy grid, and non-renewable energy is currently used to power most of the world. Only recently due to global warming and other problems such as economics, renewable energy has recently become needed more than ever. Figure 5.1 below illustrates the Cisco traffic rate exists in the data center.

The information end-users are getting is at 14% by 2020. According to Cisco's report, the volume of cloud computing workloads (74%) will be only of Software-as-a-Service (SaaS) in 2020. At the same time, Infrastructure-as-a-Service (IaaS) reduces market share from 26% in 2015 to just 17% by 2020. Platform-as-a-Service (PaaS) will also drop from 9% to 8% by 2020. Cloud computing is systematically expressed as a device mechanism where mobile users can access the services per their demand through a browser or other web-based tools [2]. Cloud computing is a structured way to allocate

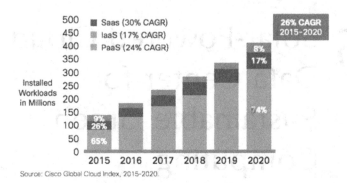

FIGURE 5.1 The Consumer Workload Is on the Rise from 21% in 2015 to 28% by 2020.

a convenient, on-demand network access of shared resources (e.g., network, servers, storage, applications, and services) that can be rapidly provisioned and released with minimal supervision or service provider interaction. In general, energy consumption can be efficiently achieved through minimal energy consumption from renewable energy resources. Obviously, its significant reward includes cost, energy conservation, and environmental guard. Mobile cloud computing has various forms like server infrastructures, networks, and mobile devices. Researchers must forfeit a scrupulous interest in energy consumption in MCC, allowing large-scale computing, network computing, and mobile users. The cloud gained popularity as companies gained a better understanding of its services and usefulness. The cloud provides three services and is getting popular in small businesses, allowing them to access application software over an internet connection without investing in software and hardware. SaaS will grow in the swift trend of the digitalized world. This increase in the market segment is growing up to $116 billion next year due to the enhancement in growing subscription-based software (Table 5.1). The second-largest market segment is the infrastructure services of cloud (i.e. IaaS), with $50 billion in 2020. Every year, IaaS emerges a 24% that yields maximum development in all market segments. This growth is attributed to the hassle of modern applications and workloads. It requires an infrastructure where traditional data centers could not apply.

According to Gartner, the three major areas of cloud computing are the most essential global CIOs that will yield maximum investment in the coming years. As organizations enlarge their trust in cloud technologies, digital IT technologies are stepping up to squeeze cloud-built software and displace existing digital resources.

Cloud services are progressively becoming more sophisticated and aggressive. Actually, there is a peak growth rate of 60% of organizations that exploit IT cloud service. These wills double the percentage of cloud service utilization rates in organizations. Cloud-resident capabilities, IT services, multi and hybrid cloud encompass a varied cloud ecosystem that will be vital for product managers. The demand for intentional cloud service results in a shift towards the digital business system.

Several definitions of cloud computing are represented as simple and most prescribed ones are suitable in the transfer of information from one device to

TABLE 5.1
Worldwide Public Cloud Service Revenue Forecast (Billions of US Dollars)

	2018	2019	2020	2021	2022
Cloud Business Process Services (BPaaS)	41.7	43.7	46.9	50.2	53.8
Cloud Application Infrastructure Services (PaaS)	26.4	32.2	39.7	48.3	58.0
Cloud Application Services (SaaS)	85.7	99.5	116.0	133.0	151.1
Cloud Management and Security Services	10.5	12.0	13.8	15.7	17.6
Cloud System Infrastructure Services (IaaS)	32.4	40.3	50.0	61.3	74.1
Total Market	**196.7**	**227.8**	**266.4**	**308.5**	**354.6**

Source: Gartner (November 2019)
BPaaS = business process as a service; IaaS = infrastructure as a service; PaaS = platform as a service; SaaS = software as a service
Note: Totals may not add up due to rounding off.

another. It is pretty evident in everyday life as it frequently used on mobile devices and internet-based applications such as Google.

As seen in Figure 5.2 below, computer infrastructures are present such as storage, a control node, and a client computer—all these electronic devices use a significant amount of energy.

Cloud computing takes a large amount of energy while executing tasks. The requirement of green energy technologies is to improve electricity production in an efficient manner by using renewable sources. At the same time, power generations are spirited down the cost of green energy systems. For example, the photovoltaic (PV) solar panel electricity is calculated to triple in 2030 effectively. The cost per watt produced by the PV solar panel is expected to halve as such if the current value is evaluated [2]. Therefore, smart green energy will be able to produce energy in future mobile cloud computing. Piro et al. [3] estimated that CO_2 emission hoard the cost of green power on various network scenarios.

It shows that powering a cloud data center with smart green energy can generate sustainable and eco-friendly solutions. Hassan et al. [4] promote the classic scenarios and the goals of renewable energy in a mobile cloud environment. This paper is organized as follows: Section 2 represents the basics of green computing and mobile computing. Section 3 presents an investigation of energy consumption cloud data centers. Section 4 explains the methodology using a solar PV generation mode and a small-scale routing algorithm. Section 5 covers the results and discussions. Finally, the conclusions and remarks are given in Section 6.

2 BACKGROUND

Cloud computing works using the following tiers:
IaaS, PaaS, SaaS

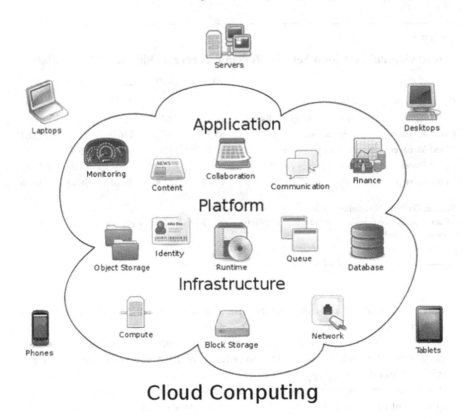

FIGURE 5.2 A Simple Diagram as to How Cloud Computing Works.

The bottom layer includes infrastructure-as-a-service (IaaS). By using IaaS, users can execute the applications on machines owned and configured by a third party. IaaS creates the hardware and software of an entire system from server to storage and networks [5].

The next layer is platform-as-a-service (PaaS). The main purpose of PaaS is to maintain the whole life cycle of the building and distribute services over the web. PaaS afford services to organize, test, host and manage applications within the same integrated development environment.

The larger observable layer to end-users is the software-as-a-service (SaaS) located on top of the PaaS layer. With SaaS, users access services through a web browser without bothering about the hardware and software details in the implementation part of the system.

These three layers are fundamental to the overall functionality of cloud computing—without them, cloud computing will be useless against other more involved internet storage methods. Cloud computing needs the development of a more efficient cloud infrastructure to reduce the overall stress of the energy consumed. These energy demands can be made by implementing energy-saving techniques, improving the functionality of computer infrastructure, and finding alternative energy means such as renewable energy. Sustainable green computing provides an appliance of both efficient energy measures and renewable energy resources to lower energy consumption and carbon footprint [6,7]. Power is generated from fuels such as coal, gas, and oil which produce a significant amount of CO_2 emissions. Smart green energy is generated by applying renewable resources like biomass, wind, and the sun which grades in almost zero carbon dioxide emissions [8]. The issues of green energy are addressed by load balancing and workload migration techniques with cloud data centers [9,10]. Moreover, the grid is designed by the power supply technique of both static network resources and dynamic renewable energy resources. This is essential to swear 100% availability of cloud services [11]. However, copious renewable energy resources are frequently located away from commercial CDC sites. Consequently, portable mobile cloud data centers are designed based on the customer and utilized in the recovery of CDC nodes close to renewable energy resources [12].

They have developed ETSA—Energy-efficient Task Scheduling Algorithm—to discuss the hitch of task scheduling. The predictable ETSA gets into account the complete operation of a task with high resource utilization and adopts the normalization procedure in making scheduling decisions. The proposed ETSA measures the total length of the schedule (that is, when all the jobs have finished processing), which results in an optimization job scheduling. The experimental results are compared with modern algorithms like round-robin, dynamic cloud scheduling, and energy-aware task scheduling. The proposed ETSA provides a well-designed operation with high energy efficiency and makespan [13]. Conversely, task scheduling is a process of reducing execution time with an efficient utilization of active resources. Modern studies reported that tasks are consigned to the virtual machines (VMs) based on their utilization value without considering their processing time. On the other hand, the task processing time is also an equivalent important criterion—it proposes the various scheduling criteria for job scheduling tasks that include both the execution time of the tasks and its resource utilization. They perform meticulous simulations by using arbitrary datasets. The results are, in turn, compared with current energy-aware task scheduling algorithms, namely random and Max-Util. The proposed algorithms progress concerns with 10% less energy consumption than the random algorithm and 5% less consumption than the Max-Util algorithm [14]. In a data entry, computer resources such as processor, memory, file, data, and I/O devices are the major components that play a vital role in power consumption. Yet, the processor is the basis of power consumption. In general, the utilization of the processor is relative to the load of the system [15] and the unused power consumption rate. The Energy Efficient Request-Based Virtual Machine Placement algorithm is about 70% in power consumption of a fully utilized server [16]. Accordingly, the idle servers must be switched off (or sleep mode) to minimize a significant amount of power; they may awaken whenever it resumes action or execution. This gives the maximum utilization of resources while running or executing the task [17].

A common approach to supervise data center energy consumption includes four main steps (see Figure 5.3): feature extraction, building model, validation of the model, and implementation of the model to real-time system.

- Feature extraction: To diminish the energy use of a data center, it needs to determine the energy consumption of its components [18] and categorize them based on the utilization rate of the energy.
- Model construction: Next, selected input features are applied in the construction of a model by using various techniques like regression, machine learning, etc. One of the keys to the problems faced in this step is that certain important system parameters—like the power consumption of a particular component in a data center—cannot be measured directly. Typical study techniques may not fabricate accurate results in such conditions and machine learning techniques may effort better. The effect of this step is a power model.
- Validation: Next, the model requirements are certified for its suitability for intended purposes.
- Model practice: Finally, the known model can be applied to predict the system's energy consumption. It can be used to develop a smart energy system. For example, by incorporating the model into various techniques like energy-aware scheduling [19,20], dynamic voltage frequency scaling (DVFS) [21–23], resource virtualization [24], civilizing the algorithms worn by the applications [25], switching to low-power states [26], power capping [27], or even completely shutting down unused servers, etc. to develop high energy data centers. Conversely, we make a note that an energy model is not eternity required for energy consumption prediction.

3 ENERGY CONSUMPTION ANALYSIS

The current overall power consumption of computer infrastructure remains unknown, but the users can be estimated. The analysis is that more than 10% of the world's energy use is consumed by the IT world—excluding machines that run

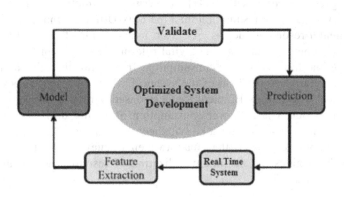

FIGURE 5.3 Optimization Model Cycle.

computer code. It needs to analyze the overall functioning of the infrastructure of equipment to reduce the carbon footprint it leaves behind and improve the outcome. On average, most of the power consumed is made by non-renewable energy substances such as coal and oil. This requires reduction to avoid a fundamental problem such as global warming—with its effects slowly hurting the earth and its surrounding areas.

Mass computer infrastructure eats a large load of current. The power equation is as follows:

$$\text{Power} = \text{Work}/\text{Time}$$

where work is any process a computer does, and time is how long it takes to process the code/functioning. The calculated amount of energy needed to power a computer peripheral depends on the amount of work the system of equipment does—in the case of cloud computing, it uses a significant amount of energy to keep the work in a stable condition and improve the output of the different services mobile cloud computing initiatives use.

As Figure 5.4 represents, every step in this process keeps 4-11% of the available power, making severe losses and additional hardware expense at every stage. When inefficiencies on both sides of the electrical zones are taken into attention, around 2 out of every 3 W of the computer equipment power are used outside the computer chips. When catching into account cooling needs, only 12% of the energy in the coal-fired power plant is used in the final computer processing. Alternatively, the renewable energy concept based on a small data center is comparatively straightforward. Figure 5.4b shows the installation of data servers into robust and transportable cabinets placed next to the renewable electricity generation site, the data server sealed to a consent outdoor facility. It is purely self-sustained with power input, network connections, and a connection for cooling and waste heat recovery. Depending on the site planning and data center design, external cooling is not necessary. Component failure within the data center permitted before overall data center performance degenerates to an agreed level. When the crash is reached, the entire data center is replaced. Thus, the renewable power plant is applied directly located in the data center.

This reduces fewer losses. Energy losses from transmission in the conventional energy access are fully usable renewable power resources. The cooling equipment combined with the small data center model and its efficiency doubled that of the traditional approach. We expected cooling as low as 20% of the mini data center's entire power load as shown in Figure 5.4b. Hence, cooling waste energy is reduced by half compared to the conventional case (Figure 5.4a). The virtual green data center access reduces cost, energy consumption, and adverse environmental impacts. Such data center depends on diverse forms of renewable energy sources that maximize execution up-time. Small capacity of mini data centers can remain connected to the power grid, which increases accuracy where transmission is extent. Small-scale battery storage, high-tech turbines, and ultracapacitors were used for continuous power delivery and extended the execution of the day.

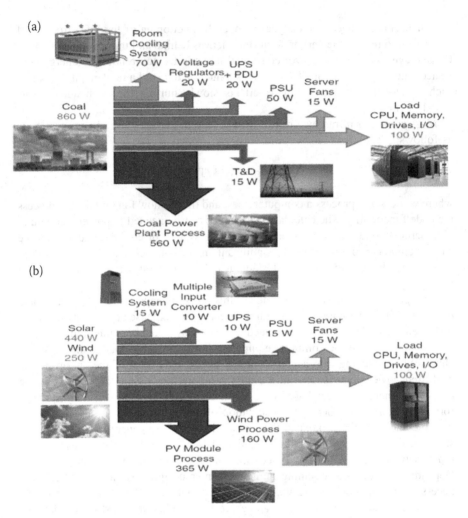

FIGURE 5.4 Brown Energy versus Renewable Power Used in Data Center Electricity Utilization. (a) Energy Used in a Conventional Data Center Approach; 860 W Energy versus Renewable of Equivalent Coal Power Is Required for a 100 W Server Load and (b) Proposed Access Based on Renewable Power Used in Data Centers; A Combined 675 W of Equivalent Power from Renewable Sources Is Needed for the Same 100 W Server Load.

Overall, improvements are expected in vital equivalent energy in this case of renewable sources with no fuel costs and the added advantage that no greenhouse gases generated, and high transmission and distribution lines are avoided. In fact, most of the energy equipment does not translate into losses, as in Figure 5.4a, but in renewable resources converted into PV modules or wind generators. Infrastructure investment cost scaled-down as no bulk investments are required to transmit power over long distances. Despite this, low efficiencies in renewable energy conversion at solar panels and wind generators involve large footprints.

Solar-Powered Cloud Data Center 79

These large footprints protect large-scale renewable energy use in medium- to large-sized data centers, but small data centers allow for much more massive use of renewable. Green mobile virtual data center access is essential because of the earth's over 7 billion population, there are over 1 billion people without electricity and mobile phones, and almost 4.7 billion people without computers and the internet [28,29]. Energy consumption and mobile communications are all considered to be critical enablers for a higher quality of life; the demand for the generation of electricity is growing. The novel concept proposed here enables widespread access to information and mobile green virtual computing as an economic growth by reducing its costs and overcoming energy constraints.

Figure 5.5 explains a model of green cloud computing data centers with its function. Most requirements in future technology are the sustainability concept of the data center. With the outlet, this is the first examination of a sustainable cloud data center that covers all the key factors of sustainability concepts and green energy in the mobile cloud. Renewable energy schemes in CDCs are classified per workload scheduling approach. The dynamic load balancing technique is preceded by the renewable approach and the server power check. The dynamic load balancing method focuses on balancing the workload within a CDC. In the renewable advancement, workload is a shift among the CDCs. The workload scheduling approach of power check is applied to voltage scale and transitions can be done by renewable power with CDCs. These techniques are sub-classified into various sources of energy and locality basis of the CDC. Depending absolutely on the renewable source, it can be directed to the power outage and subsequent services. Therefore, most CDCs are classified into a mixture of renewable power of three methods like (1) CDC with renewable power

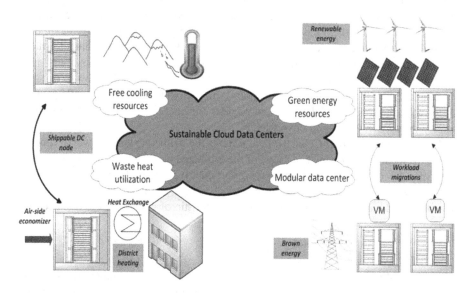

FIGURE 5.5 Sustainable Green Energy Cloud Data Centre Model.

generation plant, (2) on-site autonomous renewable power station, and (3) purchase of renewable energies from vendors with terms and conditions of purchase agreements and certificates [30,31]. The advantage of this technique is the minimal power transmission losses.

4 METHODOLOGY

4.1 CLOUD DATA CENTERS SOLAR PV MODEL

This paper focused on a small-scale data center with solar energy based on renewable resources. Several modern technologies come through, like low-power computing platforms, energy-aware cloud systems, and DC power distribution. It utilizes solar energy to generate power such as a cloud data center shown in Figure 5.6. This solar photovoltaic power generation entails mobile cloud computing and storage space nearer to the user, augment the security of data, diminish energy consumption costs, and keep a green environment. The current trend is for Google—with vast computer resources such as servers—to provide a cleaner way to generate power and replace the functioning of non-renewable energy. Renewable energy is a great alternative as this is clean energy. It includes solar power, hydro, wind, and wave technology to produce energy. The methodology is simple and, thus, improve the overall output of cloud computing peripherals by enhancing the overall power network and converting the existing grid into renewable energy—this will decrease the carbon footprint of burning fossil fuels. The method here is to optimize the output of computing related to the cloud and to maximize the energy used by improving the way the power is used. The power can be generated using renewable energy resources. Non-renewable energy resources are ineffective and cause major problems related to the environment. Large computer rooms also need cooler temperatures compared to the surrounding environment, as this is vital for the processing speed and the output generated. Because of this, it needs to reduce the amount of energy used and improve maximum output. The solar photovoltaic (PV) generation model deals with its characteristics and the equations of current-voltage—this is related to solar cell parameters and environmental factors adopted in differential algorithm [32]. Solar power can be calculated from the maximal power point (mpp) extracted from the PV panel. Differently, the load-scheduling system predicts the energy cohort as a function inversely proportional to cloud coverage.

$Ep(t) = B(t) (1 - Cloud\ Cover)$, where $Ep(t)$ is the deliberate solar power at time t, $B(t)$ is the ideal solar power. The range of the estimated percentage of cloud coverage is within 0~1 (compared to sunny days). Particularly, the single diode equation is widely used to simulate available electrical power from a single PV panel. Specifically, the resulting current-voltage characteristic of a PV panel is denoted as,

$$I = I_{ph} - I_0 \cdot (e^{v + i \cdot R_s / n_s V} th - 1) - v + I \cdot R_s / R_{sh} \tag{1}$$

where I_{ph} is the photogenerated current, while $I0$ is the dark saturation current concerned with the ambient weather pattern. Furthermore, the single-diode model

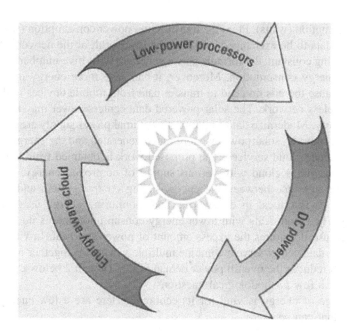

FIGURE 5.6 A New Generation of Green Solar Powered Cloud Computing Data Centers.

considers both the series and parallel resistance as Rs and Rsh, respectively. Vth is the junction thermal voltage:

$$V_{th} = k \cdot T / q$$

where k is Boltzmann's constant, q is the charge of the electron, and T is the ambient temperature.

ns is the number of the solar cells in the PV panel connected in series (e.g., $ns = 72$ in BP-MSX 120 panels).

The solar power makes use of meteorological data from the Measurement and Instrumentation Data Center (MIDC) [33] of the National Renewable Energy Laboratory. Presently, most cloud data centers have inefficient power exchange. Power is generated by alternating current (AC) that passes through multiple conversions among AC and direct current (DC), resulting in up to 30% energy loss. In the future, DC-solar Photovoltaic-based power distribution systems may improve the renewable energy utilization rate and minimize total power usage. The amount of power is saved by applying the formula,

$$P_c X (C/M) - P_i X (C/S) - P_{tr} X (D/B) \qquad (2)$$

where C is the number of offloaded computations, M is the speed of the mobile device in terms of instructions per second, S is the speed of the cloud server (instructions/second), Pc is the computational power consumption (watts), Pi is idle power consumption (watts), Ptr is the transmission power consumption (watts), D is the bytes of data to be exchanged, and B is the bandwidth of the network. With all Pc, Pi, Ptr being constant, the formula above generates a positive number, indicating minimized energy consumption. Moreover, it can save more energy in the cloud platform because there is no need to transfer data from mobile devices to the cloud over the wireless network. The solar-powered data centers power may take part in an efficient way. Modernize data centers with minimal power supply are around 25-28% [34]. In summer, solar power is more much generated and the energy-sensitive cloud will reduce cloud services. The proposed work is obtained from the existing one and it facilitates cloud with various supply of controlled energy. Figure 5.7 describes the interface between the energy-saving control system and the cloud environment. It is critical in discovering and monitoring energy usage in cloud computing. The result deals with lower energy consumption that is no longer reasonable because it requires the excess amount of power and heat. In virtualization technologies, data center servers combine multiple workloads together into a single server which reduces the overall power demand [35]. Table 5.2 below explains and helps answer a few methodological questions:

The change of energy is vital for its economy. Here are a few energy-saving techniques that can be used:

1. Improve the surrounding areas by developing more evolved data centers—this means improving the output and energy consumed.
2. Implementing greener energy solutions such as solar and wind technology to generate the energy needed to keep large computer systems operational.
3. To maximize the output and improve the computer processing output.

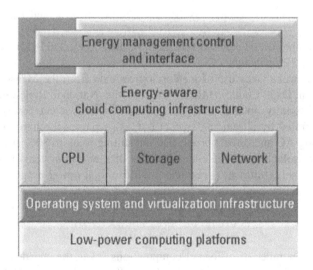

FIGURE 5.7 The Recognition of Energy Control System and the Cloud Environment Versus the Usage of Energy in Cloud Platform.

TABLE 5.2
Methodological Motivations

Questions	Motivation
Why is energy efficiency important?	Energy efficiency is important to improve overall understanding of the way energy is consumed and to preserve the way energy is used.
Renewable or non-renewable energy?	Renewable energy is more viable and reliable even though it is costly. Non-renewable energy resources release harmful gases into the environment which is not excellent.
How is cloud computing beneficial in energy efficiency?	It is advantageous as computing with clouds has surfaced as a useful paradigm in handling the energy problem and dealing with the growing need for power. There are many different techniques implemented in cloud computing to help curb the rising power consumption in the overall cloud computing sector.
Methods to use?	Implement renewable energy schemes around big data centers to help curb the need for the energy required by the overall power grid.
IT industry roles?	The role of IT industries is simple — they can create evolved techniques to help reduce the overall energy consumed by their big data equipment and improve their energy efficiency technologies.
The role of cloud computing?	Cloud computing play an excellent part in the world of technology. We can implement larger drives to improve the way energy is harnessed and consumed.

4. Implementing cruel laws and regulations on IT companies to apply green energy to help decrease the load on the already-overloaded electricity grid.

Virtualization technologies can be supported by cloud data centers in two ways. First, combine several workloads into a smaller number of servers, hoist operation levels, and improve energy efficiency. Then, the virtualized cloud data center decides to select the work among them for execution and allocate the requested resources. This happens in a more efficient manner with greater effort because cloud virtualization facilitates resource portion energetically within a single server. It creates to move workloads from server to server [36].

The minimum shortest path routing provides less-delay and maximum number of completion of jobs during data transmission. It selects the least number of costs in the route. The minimum shortest path routing is not only energy-aware but could also lead to fast execution time. The results are analyzed and evaluated by the power consumption in a cloud environment by means of the impact of Solar Photo Voltaic System configurations. Solar DC power supply in the data center removes the requirement of inefficient AC-DC transformers in power supplies, which decreases not only the cooling costs but also the total data center footprint [37]. The solar green power is calculated approximately in an efficient data center as 42% of the total power is used by air conditioning systems [38]. The first explanation covers an implementation of the process set up or how solar power is the suitable and sufficient energy providing. Free Launch [39] is a mobile cloud data center for structural design to the sustainable cloud data center elements. Free launch is based on three principles: (a) utilization of on-site solar energy through remote CDCs, (b) devoted high-speed network connectivity between two CDC centers, and (c) virtual machine-based workload movements. In virtualization, a mobile data cloud center with solar energy is a key to enable sustainable CDCs. The advancement of self-adaptive systems is to meet the requirements for autonomic computing [40]. It needs self-adaptive software systems that ensure all energy aspects with less attention [41], as that can be reasonable to the challenges that meet production with a suspicious operating environment [42]. The requirement of energy efficiency of the cloud environment is described significantly by Djemame et al. [43].

4.2 SMALL SCALE ROUTING ALGORITHM

This algorithm minimizes energy consumption instead of using solar energy, thus, increasing effective performance upgrade. The pseudocode has the following steps as given below:

1. Create the network connectivity from cloud providers to data centers.
2. Initialize the source of solar energy and power reading info.
3. Send the request from the cloud provider to the data center.
4. Based on the sampling requirement, data return to the cloud provider.
5. Get the power monitor log for a period.
6. Set the range of solar power and get the routing path.
7. Base the routing and estimate the solar power consumption.
8. Get the optimal green power path.
9. Get the measured route table.
10. Power on/off.
11. Return the pathway.

Small scale routing algorithm
   Small Scale Routing (M,r,bw).
   Input:
   Input: SSR (routing matrix), T (traffic matrix)
   Output:
   P(routing path)
   Set M (V, E, L) Cloud data center network Structure and link;
   r: Flow of Request,
   bw: bandwidth for the flow,
   W (link weight),
   U R(usersrequirementsmatrix),

C max(link capacity)
Invoke algorithm OLS to find out idle nodes [new W, sl = OLS(R, W);
new W shows the new link weights
sl shows the number of sleeping links
m = Constant;
set the number of users' requirements for i = 1, 2, ..., m do
identify the Dijkstra algorithm
Path = Dijkstra(W, Source, Destination);
the weighted adaptive shortest path first if Path = 0 do k = length(Path);
call the Get_power algorithm
P: Path for flow
Path X: Get_ all_ Path(M, r)
Path Y: Get_ available_ Path(Path X; bw)
if Path Y is empty then return Zero
else
for Path Y in Path X
I←0
M'=Add_ Network Utilization(M, Path Y, bw)
Pow [i]= Get_ Pow(M')
I←I+1
end for
index← Get_ Index (min(pow))
return Path X[index]
end if

For each stepping flow process, the routing path is found out. All pairs of paths are useful in generating a subset of the network. The small-scale routing algorithm is a simple solution to power minimization. Each step flow, the small-scale routing algorithm verifies all possible paths with adequate capacity and traverses all the pairs of paths—choosing the shortest path with the slightest power consumption. The evaluation of power consumption depends on the power measurement data derived from power models.

Table 5.3 shows the iteration 1 energy level. Cloud providers execute the first iteration at an idle state and becomes fully active while consuming power. Here, the power consumption is measured in a well-known way where all interfaces are powered; the power consumption of a crossing point can be identified.

The routing algorithm is used to curtail the total energy consumption in the network by concentrating on methods for reduction. In this paper, the proposed method seeks to look at the problem of balancing energy consumption by exploiting the nodes that have enough residual energy.

Detection of Route

The route detection segments include two sub-components: the first one is the next-hop selection and the next is network measurement. In the first component, the set of proper next-hops are found by applying criteria, which offer optimum data relay and balances the transmission load with consistent energy consumption. During routing decision, the proposed protocol make use of fitness function to determine Forwarder Point (FP) of node i, which integrate hop_count

TABLE 5.3
Iteration 1 Information

Small Scale Data Center

Controller IP	Virtual machine
Controller port	8086
Power consumption	1359.77 kWh
Energy Source ID	
Solar power generation	1360 kWh
Solar power price	Rs. 3.5 /kWh
CO_2 emission rate	0.010 kg/kWh

h_count_i, residual energy e_i, and Weighted Round Trip Time $W_{RTTi,j}$ factors, as shown below:

$$FP = w_1 * e_i + w_2 * \frac{1}{h\_count_i} + w_3 * \frac{1}{W_{RTTi,j}}$$

The computed value of FP and all the three factor's residual energy, hop-count, and weighted RTT are normalized in the range of (0.0, 1.0). The coefficients w_1, w_2, w_3 represent the weighting factors—the residual energy, hop-count, and weighted RTT. Consequently, the node that optimizes the amalgamated routing function in terms of energy is hop-count. The weighted RTT is chosen as next-hop. It might be a case that more than one next-hop has the same FP values. In such a case, Node ID breaks the ties, and the source node keeps only a single entry in its neighbor table. During data forward, the integration of energy factors in routing decision significant impacts on network lifetime with route strength. Additionally, maximum network throughput is attained due to the assortment of least RTT next-hop. Moreover, it incorporates the slightest number of nodes during routing; the shortest routes are devised based on hop-counts. To construct a neighbor table, each source node i broadcasts RREQ in its transmission range and on receiving, store the residual energy and number of hops information. In addition, each node i determines RTT for a beacon messages X_i at node j as shown in Eq (5.1)

$$RTT_{i,j} = \sum (R_x - T_x) \qquad (5.1)$$

Map Reading Routing Algorithms1 directs the phases of the proposed RCER protocol.
MRR Algorithm: clusters formation, next-hop selection, and network measurement phases

1. Begin
2. procedure CLUSTERS
3. Determine the set of centroidal based on nodes location
4. for (I = 1; I < = Nodes; I + +)
5. do

6. If $node_{energy} > optimistic_{threshold}$
7. setlist of candidates []
8. end if
9. end for
10. for (I = 1; I <= list of candidates; I + +)
11. compute the node centrality of node $_i$
12. set the max(closeness node $_i$) as CH $_i$
13. end for
14. end procedure
15. procedure **route detection**
16. for each $node_i \in [1:M]$
17. do
18. call optimized data_ routing ()
19. while (y! = destination)
20. $node_i$ sets next-hop by using highest FP_i
21. $F P_i$ Reply to y
22. $y = F P_i$
23. if y_i. latency epoch is high
24. call route restore ()
25. end if
26. end while
27. end for
28. end procedure
29. Procedure nodes_ status
30. if (CH_i. energy $<\mu^*$ $energy_{init}$)
31. initiate re-election ()
32. if (Δt expired) then
33. initiate re-election ()
34. end if
35. end if
36. end procedure
37. End

4.3 MRR Analysis

The quality of the proposed MRR protocol is compared to active schemes.

1. The MRR protocol is intended for heterogeneous WSNs based on the cluster-based solution capable of energy efficiency with reliable routing.
2. The clusters formation is achieved based on nodes region, which produces the generation of more upright clusters with negligible energy consumption.
3. To achieve optimal routing choice, the fitness function is used based on multiple factors. It balances energy consumption and diminishes routing overheads, as only neighboring nodes take part in routes construction procedure.

TABLE 5.4
Power Consumption by the Selected Servers at Different Load Levels in Watt

Servers	0%	10%	20%	30%	40%	50%	60%	70%	80%	90%	100%
Dell Inc Power Edge M620	688	1151	1322	1494	1671	1848	2061	2289	2499	2765	3239
IBM NeXt Scale nx360 M4	550	873	999	1123	1251	1380	1525	1673	1887	2116	2404

5 RESULTS AND DISCUSSIONS

These are the two selected server configurations with multi-core CPUs. The configuration and power consumption characteristics of the selected servers are shown in Table 5.4. So, the build model depicts that total data center power(t) = server power consumption(t) X PUE value.

Moreover, we define some feature that classifies the above-mentioned works for a better understanding of the problem that we want to tackle.

Green energy consumption in the data center: By default, we mean that an application is aware of the predictability of green energy and changes its execution time or makespan accordingly. Figure 5.8 expresses our goal to adopt an interactive application based on the aforementioned events.

Outlay optimization: Most data center research is rigorous on how to reduce electricity-related costs. Solar-powered tied energy rates can be compact by using smart green energy, whereas efforts have been made to develop location and price assortment for further reduction.

Recital guarantee via edition: Performance is the primary criterion for any cloud application, but it can vary depending on its nature. For example, abiding by the target is a primary requirement for batch application. On the other hand, latency and ease of use are the key concern for interactive application.

In a way, these approaches tried to make an application by using smart renewable green energy technology. In literature, works related to interactive cloud application mostly emphasize to guarantee concert by doing architectural change in run-time to allow more requests that preserve shortage of resources (functional adaptation) or adding more resources to satisfy demand spikes (non-functional adaptation), or formulating them together. This policy actually opens the door for creating green energy adaptively to the interactive cloud application. However, literature studies still lack the mechanism of how the interactive application can take advantage of the smart green energy to re-configure itself. It makes changes to reduce brown energy where it satisfies all performance requirements.

To overcome the problem of creating green energy knowledge in mobile cloud applications, we proposed several application controllers that make a trade-off between brown and green energy consumption to acknowledge the idea of adapting the smart mobile applications with green energy. Naturally, mobile applications will

Solar-Powered Cloud Data Center

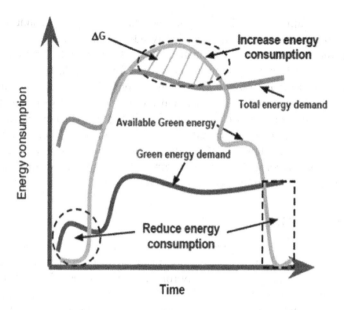

FIGURE 5.8 Green Energy Adaptive Interactive SaaS Application.

require the computing resources of mobile devices as well as the servers in a data center in the cloud. Therefore, the mobile cloud service providers have to provide both radio and computing resources and jointly optimize them to attain the maximum revenue. The proposed cloud structure deals with energy efficiency at all layers and throughout the complete cloud application.

The major discussion is that it needs to address the integration of solar power with cloud data centers. It was explained in the part of the energy consumption of cloud infrastructure. Mainly, it observes power consumption by attaining two different scenarios:

- Acquire the data from the cloud data center that uses a smartphone network, the internet, and the cloud computing data center.
- Acquire the data from the mobile cloud data center that uses a smartphone network, MCC data center, the internet, and CC data center.

The first endeavor is to minimize energy usage and maximize energy efficiency in computing devices and applications. Energy-saving requires efforts of minimal power consumption, which is achieved through high power efficiency with power consumption from conventional energy sources. Mobile cloud maintains the implementation of mobile client applications without paying much attention to efficient mobile resource management.

Shortest path routing is followed by a low delay and high throughput of data transmission. It selects a routing path with the least number of lengths or weights. Shortest path routing is not power-aware, but it could lead to fast execution to complete the tasks. Typical examples consist of the implementation of an application, scanning for Wi-Fi network, and matching email account during the night when mobile users are sleeping. Many ad-hoc techniques are accessible to reduce power drainage. These are inadequate to save enough energy on mobile devices. Naturally, it is very essential to draw the following things: cost reduction, energy conservation, and environmental protection. Mobile clouds include computing server infrastructures, networks, and mobile devices. An MCC data center consists of all the elements of a CC data center but to a smaller extent. Energy consumed by smartphones is the same as obtaining data download types in networks. Energy consumed by the MCC data center would be self-effacing along with their needs. Energy consumed by the internet is the same for both data centers, transferring the content to the MCC data center.

6 CONCLUSIONS

This paper depicts the solar panel energy power consumption in cloud environments and motivates them to save energy. In this respect, it also described power consumption in a smart green cloud environment and mobile offloading of energy-saving formulas. Furthermore, it provides a generic solar PV Power generation model with two states of idle server and active server states. This also discusses the evaluation of a small-scale routing algorithm strategy based on an initial prototype of energy consumption. This prototype demonstrates the computation of the energy, cost, and sustainability information for cloud providers. The small-scale routing algorithm is used to find the most energy-efficient paths among the shortest paths. This research results are very important to develop potential energy legislation and management. This mechanism is used to minimize power consumption with greater system performance in cloud environments. Future work will investigate several new cloud environments for mobile users and propose new renewable energy power methodology used to decrease CO_2 emissions. The solar-powered cloud data centers can be the future generation of intensification in the global internet communications foundation. The various research challenges are gear for prolific integration of renewable energy to CDCs. The most important dispute that needs to be addressed with respect to the incorporation of renewable energy to CDCs is the intermittency of renewable energy. The hybrid framework provides the architecture of renewable power supply. The design is focused to ensure the availability of cloud services and non-violation of SLAs. The system development life cycle is based on the review and competence planning of CDCs in terms of energy cost. The business models are required for sustainable CDCs. Although work has been completed, it does not cover up all aspects of the sustainable CDC system. Precisely, the prediction model of renewable energy generation and sustainability of the CDC is serious — these are based on weather forecasts and predictions. Similarly, location-aware data center load balancing system also needs to be investigated. Thus, researchers will integrate energy costs with a new model by differing environmental impacts and minimize the total energy cost.

REFERENCES

1. Mell, P. and Grance, T. (2009 The NIST definition of cloud computing, *NIST*, 53 (6), pp. 1–7.
2. Manash, E. B. K. and Rani, T. U. (2015). Cloud computing – A potential area for research. *International Journal of Computer Trends and Technology*, 25, pp. 10–11.
3. Piro, G., Miozzo, M., Forte, G., Baldo, N., Grieco, L.A., Boggia, G., Dini, P. (2013 Hetnets powered by renewable energy sources: Sustainable next-generation cellular networks. *IEEE Internet Computing*, 17(1), pp. 32–39.
4. Hassan, H. A. H., Nuaymi, L., and Pelov, A. (2013). Classification of renewable energy scenarios and objectives for cellular networks. In Proceedings of the IEEE PIMRC '13, London, U.K.
5. https://apprenda.com/library/architecture/the-three-tier-model-of-cloud-computing/.
6. Masanet, E., Shehabi, A., and Koomey, J (2013). Characteristics of low-carbon data centers. *Nature Climate Change*, 3(7), pp. 627–630.
7. Uchechukwu, A., Li, K., ad Shen, Y. (2012). Improving cloud computing energy efficiency. In Cloud Computing Congress (APCloudCC) 2012, IEEE Asia Pacific (pp. 53–58). IEEE.
8. Deng, W., Liu, F., Jin, H., B., Li, Li and D. (2014). Harnessing renewable energy in cloud data centers: Opportunities and challenges. *IEEE Network*, 28(1), pp. 48–55.
9. Qureshi, A. (2010). *Power-demand routing in massive geo-distributed systems* (Ph.D. Dissertation). Massachusetts Institute of Technology.
10. Mustafa, S., Nazir, B., Hayat, A., and Madani, S.A. (2015). Resource management in cloud computing: Taxonomy, prospects, and challenges. *Computer & Electrical Engineering*, 47, pp. 186–203.
11. Oro´, E., Depoorter, V., Garcia, A., Salom, J. (2015 Energy efficiency and renewable energy integration in data centers. Strategies and modelling review, *Renewable and Sustainable Energy Reviews*, 42, pp. 429–445.
12. I., Goiri, W., Katsak, K., Le, T.D., Nguyen, R., Bianchini, 2013 In: Sigarcha, C.M., editor. Parasol and greens witch: Managing data centers powered by renewable energy.Computer Architecture News, 41, pp. 51–64.
13. Panda, S. K. and Jana, P. K. (2019). An energy-efficient task scheduling algorithm for heterogeneous cloud computing systems. *Cluster Computer*, 22(2), pp. 509–527.
14. S. K. Panda and P. K. Jana, (2015). An efficient task consolidation algorithm for cloud computing systems. In: Bjørner N., Prasad S., Parida L. (eds.) Distributed Computing and Internet Technology. ICDCIT 2016. Lecture Notes in Computer Science (vol. 9581, pp. 61–74). Springer, Cham.
15. Beloglazov, A., Abawajy, J., and Buyya, R. (2012). Energy-aware resource allocation heuristics for efficient management of data centers for cloud computing. *Future Generation Computer Systems*, 28, pp. 755–768.
16. Kusic, D., Kephart, J. O., Hanson, J. E., Kandasamy, N. (2008). Power and performance management of virtualized computing environments via lookahead control. In: *International Conference on Autonomic Computing* (pp. 2–12). IEEE.
17. Panda, S. K. and Jana, P. K. (2016). An efficient request-based virtual machine placement algorithm for cloud computing. In: Krishnan P., Radha Krishna P., Parida L. (eds) Distributed Computing and Internet Technology. ICDCIT 2017. Lecture Notes in Computer Science. Springer (vol. 10109, pp. 129–143). Cham.
18. Brown, D. J. and Reams, C. (2010). Toward energy-efficient computing. *Queue*, 8(2), pp. 30–43.
19. Bellosa, F. (2000). The benefits of event: Driven energy accounting in power-sensitive systems. In: Proceedings of the 9th Workshop on ACM SIGOPS European Workshop: Beyond the PC: New Challenges for the Operating System, ser. EW 9 (pp. 37–42). ACM, New York, NY.

20. Jing, S.-Y., Ali, S., She. K., and Zhong. Y. (2013). State-of-the-art research study for green cloud computing. *The Journal of Supercomputing*, 65(1), pp. 445–468.
21. Hotta, Y., Sato, M., Kimura, H., Matsuoka, S., Boku T., and Takahashi, D. (2006, April). Profile-based optimization of power performance by using dynamic voltage scaling on a pc cluster. In: Parallel and Distributed Processing Symposium, 2006. IPDPS 2006. 20th International (pp. 1–8).
22. Weiser, M., Welch, B., A., Demers, and Shenker, S. (1994). Scheduling for reduced cpu energy. In: Proceedings of the 1st USENIX Conference on Operating Systems Design and Implementation, ser. OSDI '94. USENIX Association, Berkeley, CA, USA.
23. Liu, H., Xu, C.-Z., Jin, H., Gong, J., Liao, X. (2011). Performance and energy modeling for live migration of virtual machines. In: Proceedings of the 20th International Symposium on High Performance Distributed Computing, ser. HPDC '11 (pp. 171–182). ACM, New York, NY.
24. Beloglazov, A. and Buyya, R. (2010, May). Energy efficient resource management in virtualized cloud data centers. In Cluster, Cloud and Grid Computing (CCGrid), 2010 10th IEEE/ACM International Conference on (pp. 826–831).
25. Feller, E., Rohr, C., Margery, D., and Morin, C. (2012, June). Energy management in iaas clouds: A holistic approach. In Cloud Computing (CLOUD), 2012 IEEE 5th International Conference on (pp. 204–212).
26. Lefurgy, C., Wang, X., and Ware, M. (2008). Power capping: A prelude to power shifting, Cluster Computing, *11*(2), pp. 183–195.
27. M. Lin, A. Wierman, L. Andrew, E. Thereska (2013, October). Dynamic right-sizing for power-proportional data centers. *Networking, IEEE/ACM Transactions on*, 21(5), pp. 1378–1391.
28. Dutta, S., Dutton, W. H., and Law, G. (2011). The new Internet world: A global perspective on freedom of expression, privacy, trust and security online. World Economic Forum.
29. Sharma, C.(2012). Global Mobile Market Update. http://www.chetansharma.com/GlobalMobileMarketUpdate2012.htm. [(accessed 9.12.12)].
30. Ren, C., Wang, D., Urgaonkar, B., and Sivasubramaniam, A. (2012). Carbon-aware energy capacity planning for datacenters. In: 2012 IEEE20th International Symposium on Modeling, Analysis & Simulation of Computer and Tele-communication Systems (MASCOTS) (pp. 391–400).
31. Goiri, I., Katsak, W., Le, K., Nguyen, T. D., and Bianchini, R.(2014). Designing and managing data enters powered by renewable energy. *IEEE Micro*, *34*(3), pp. 8–16.
32. Sudha, M. R., and Sornam, M. (2016). Differential evolutionary algorithm for energy efficiency improvement on mobile cloud computing. *International Journal of Control Theory and Applications*, 9(40) 1149–1156.
33. Measurement and Instrumentation data center. http://www.nrel.gov/midc/.
34. Bergqvist, S. (2011). Energy Efficiency Improvements using DC in Data Centers (Master's thesis). Uppsala Universitet.
35. Laura Grit, G., David, I., Aydan, Y., and Jeff, C. (2006). Virtual machine hosting for networked clusters: Building the foundations for 'autonomic' orchestration. In: Proceedings of the 2nd International Workshop on Virtualization Technology in Distributed Computing (VTDC 06) (p. 7). ACM.
36. Berl, A., Gelenbe, E., Di Girolamo, M., Giuliani, G., De Meer, H., Dang, M. Q., and Pentikousis, K. (2010). Energy-efficient cloud computing.*The Computer Journal*, *53*(7), pp. 1045–1051.
37. Ton, M., Fortenbery, B., and Tschudi, W. (2008, March). *DC power for improved data center efficiency*. Lawrence Berkeley National Laboratory.
38. Guidelines for Energy-Efficient Data Centers, white paper (2007, 16 February). The

Green Grid Consortium. www.thegreengrid.org/~/media/WhitePapers/Green_Grid_Guidelines_WP.pdf?lang=en.
39. Akoush, S., Sohan, R., Rice, A., Moore, A. W., and Hopper, A. (2011). Free lunch: Exploiting renewable energy for computing. In: Proceedings of HotOS (pp. 17–17).
40. Giese, H., Muller, H., Shaw, M., and De, R. (2008). Lemos. 10431 Abstracts collection – Software engineering for self-adaptive systems. Schloss Dagstuhl - Leibniz-Zentrum fuer Informatik.
41. Qureshi, N. and Perini, A. (2009, May). Engineering adaptive requirements. In 2009 ICSE Workshop on Software Engineering for Adaptive and Self-Managing Systems, 2009 (pp. 126–131). IEEE.
42. Camara, J. and De Lemos, R. (2012, June). Evaluation of resilience in self-adaptive systems using probabilistic model-checking. In 2012 7th International Symposium on Software Engineering for Adaptive and Self-Managing Systems (SEAMS) (pp. 53–62). IEEE.
43. Djemame, K., Armstrong, D., Kavanagh, R. E., Ferrer, A. J., Perez, D. G., Antona, D. R., Deprez, J. C., Ponsard, C., Ortiz, D., Macıas, M., Guitart, J., Lordan, F., Ejarque, J., Sirvent, R., Badia, R. M., Kammer, M., Kao, O., Agiatzidou, E., Dimakis, A., Courcoubetis, C., and Blasi, L. (2014, August). Energy efficiency embedded service lifecycle: Towards energy efficient cloud computing architecture. In the Proceedings of the Workshop on Energy Efficient Systems (EES'2014) at ICT4S (pp. 1–6). Stockholm, Sweden.

6 State-of-the-Art Energy Grid with Cognitive Behavior and Blockchain Techniques

R. Rajaguru[1], S. Praveen Kumar[1], and R. Krishna Prasanth[1]

Department of Computer Science and Engineering, Sethu Institute of Technology, Virudhunagar, Tamil Nadu

1 INTRODUCTION

Electricity is one of the essential factors for the development of societies. Energy industries were facing many challenges as they have an increase in power demand, which creates stress on conventional power plants [1]. Globally, over 80% of energy is produced by burning fossil fuels, and its contribution to greenhouse gases is two-thirds [2]; these fuels also induce solid wastes. Hence, the areas around the power plants require proper topsoil. The cost of safety precautions and the disposal of residues of these power plants were also rising, leading to an increase in production cost, which is why some industries are compromising safety measures. Using archaic technologies to operate power grids, resource utilization is ineffective [3]. But the current trends in energy consumption of the individuals, as well as the commercial organizations, were gradually increasing. Besides environmental degradation, conventional energy sources will not be sufficient to cope with the increasing energy requirements. These factors are now demanding renewable energy sources on a large scale as well as at micro-levels. The distributive architecture implementation is difficult with conventional power plant structures, and it results in single node failure. This is the main reason for massive scale outages with the delayed recovery process and it also leads to energy loss in case of distant transmission. For these reasons, renewable power sources are needed. They can be distributively installed, and energy routed effectively with minimum damage. But the uncertainty in renewable sources is not handled efficiently by traditional practice. So today, there is a need for a replacement of the conventional energy practice, and the limited natural resources consumption can be lowered with renewable sources and reduce pollution, the cost of operations—to provide a better world to the future generation.

1.1 Renewable Energy's Unreliable Nature

The industries confiding in traditional power stations are hesitant to pursue renewable power sources because of their dubious nature. This aspect may affect production and results in a considerable loss. Instead of completely committing to renewable energy, some industries take solar and windmill stations as a support system for the powerhouse. But in most of these situations, the resource utilization is not up to mark, and the maintenance makes it costly. In terms of domestic usage, the average electricity usage of a home in India is nearly 90 units per month. A 1 kW solar panel cost around Rs. 35000, which produces 3–5 units of energy per day. To support the grid, 150 Ah battery can be used, which costs around Rs. 12000. The total solar energy system for average household usage, including inverter and cables, costs approximately Rs. 9000 for a completely off-grid system. The effective utilization of this grid is required to regain the invested money hastily. These power systems were purely dependent upon environmental conditions such as daylight for solar panels and wind flow for wind turbines. Therefore, there is an elevated risk of outages. Because of this attribute, there is a need for an energy storage system. The installation cost of solar panels and wind turbines is higher but profitable in the long run. The enhancement of the grid architecture in the security and performance aspects with the latest available technologies is this secure self-restrained energy grid.

1.2 Distributed Architecture Benefits

Conventional and non-conventional power plants require an organizing structure to reduce energy wastage and disaster management. The centralized architecture is simple for construction; it is more prone to single-point failures. The grid architecture addresses some of the issues in the direct supply of energy to a more significant number of consumers from a single station. The distributed architecture resolves the single-point failure. The practice of decentralized architecture helps manage node failure; that is, if a power plant faces any planned or unplanned shutdowns, the rest of the plants can help balance the load. Distributed Generation (DG) is a supply technology that can be at or near retail capacity, enabling smart buildings and power park/premium operating districts, providing high-quality, 99.999% reliability. Combined with on-site energy storage, DG is a disruptive technology transforming centrally managed radial delivery infrastructure into a geodesic network, providing higher resilience of the power system, and enabling consumer integration into energy markets. Traditionally more associated with high-availability fuel cells, increasingly renewable solutions—most notably photovoltaic (PV) solar—are being included as part of a DG solution.

1.3 Renewable Energy Resources

Renewable energy sources implemented in distributed grid architecture make it much more sustainable [4]. Nowadays, the latest technologies and competition in the market results in reduced component costs. Solar panels are becoming the cheapest, most compact, and more reliable among other green energy sources. The prosumers are gradually developing a trend where consumers also contribute to the grid by selling the surplus electricity generated by their rooftop solar panels. There are several solar panels: monocrystalline, polycrystalline, amorphous silicon panels, and concentrated PV cells. Each type of board has its advantages and limitations with varying price ranges. Blockchain can provide convenience for peer-to-peer (P2P) energy trading that eliminates the dependency on the third party for trust [5]. The further acceleration in trend through the decreasing costs of residential-scale battery systems increases the self-consumption and self-sufficiency rate of prosumer households [6]. Smart contracts extend blockchain technology to a wide range of commercial applications [7]. The investment in renewable energy is not a sudden profitable capital; it is on a long road. By recent Greenpeace estimates, the world could save around $180 billion a year by switching 70% of the world electricity production to renewable options—this alone is an excellent economic argument in favor of renewable energy practice. Economic advantages are the main drivers in promoting solar energy. The energy transition towards a low-carbon society is high on the demand for a stable environment.

1.4 NEED FOR EFFECTIVE UTILIZATION

The adoption of renewable energy on a large scale for public and commercial demand is going to be costly. Hence, the energy grid sector requires a reformation to withstand that pressure and compensate for the investment. For that, we proposed an architecture that mainly concentrates on battery life span prolonging through charge/discharge control, reducing transmission losses of energy, and making the grid network data secure with the latest technologies.

2 RELATED WORKS

2.1 EXISTING SYSTEMS

The smart grid (SG) concepts have revolutionized the future of the conventional electric grid by making it more efficient, resilient, and reliable. Modern equipment and automation control systems have provided benefits such as quick communication, increased power quality, cheap electricity, and low operational costs. The two-way communication paradigm has enabled consumers to take control of their energy consumption and electricity bills, and the IoT automation reduces extensive human intervention with real-time energy distribution monitoring [8]. However, recent field tests have shown that wireless links in smart grid environments have high packet loss and varying link qualities because of equipment noises and obstructions [9]. Optimization techniques are required to meet the quality of service, active communication, and efficient data [10]. The architecture of the general smart energy grid (see Figure 6.1) comprises the power production unit, storage unit, and the data communicated to get stored in a standard database. The smart grid offers security, but the concept adds complexity to the existing electrical industry [11].

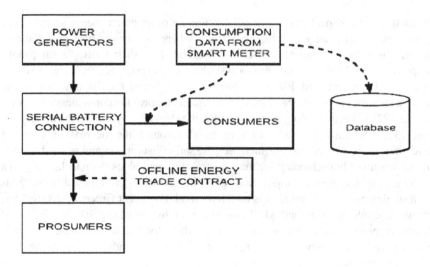

FIGURE 6.1 The Workflow of the Existing System.

2.2 Problems with Existing System

The energy grid requires an effective storage system to store energy without any wastage. There is no effective control in existing systems over energy storage to utilize battery life span completely. Therefore, it requires an engineering solution instead of merely having more storage. The static billing structure practices in the existing system are not suitable to benefit consumers with varying energy needs. Hence, appropriate billing is needed according to the individual consumption behavior and energy contribution to the grid [12]. Things such as, "how much stress does a consumer deploy on the grid?" should be considered. Maintenance should be scheduled not to interrupt high energy production hours, and the major feature expected by the customer is on-demand service that is a scalable utilization of service. The grid data are at high risk in terms of privacy and security because IoT devices brought into the grid are more prone to cyber-attacks [13]. Modifying a single data may disrupt the entire grid. The energy distribution in the existing grids is at higher levels in the infrastructure, which is not able to reduce the energy transfer loss, especially in the case of prosumers. The renewable energy sources are uncertain, so it is useful only if complete resource utilization takes place.

3 PROPOSED GRID DESIGN

3.1 Architecture

The main characteristic of networked architecture in the energy sector is distribution. The pool of interconnected battery banks acts as a mass storage unit for the architecture. Copper cables were used for power transmission all over the grid, and the data transferred through wireless communication devices. High voltage relay switches placed between transmission lines in the battery pool controls the

State-of-the-Art Energy Grid

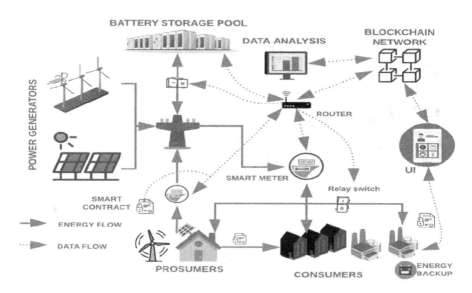

FIGURE 6.2 The Architecture of Cognitive Grid Manager.

charging/discharging operations on respective battery banks. Energy meters are connected to embedded processors with digital energy reading sensors. This smart meter setup is fixed in each consumer terminals and connected to a nearby communication hub to form a communication network. Prosumers are also provided with the same structure to manage their energy trades. The application codes are deployed on the Ethereum blockchain platform as smart contracts. With that, an Interplanetary File System (IPFS) node is hosted over the communication network where the grid data are routed. Load forecasting model and clustering model deployed through API. The web application serves as a User Interface; through that, the administrator controls grid operations, and customers monitor the service usage. The complete structure of the energy grid is represented in graphical form (see Figure 6.2).

3.2 SMART ENERGY DISTRIBUTION

A self-restrained energy grid requires a capable storage unit and flow controlling devices with communicating nature. Then, the data should be processed and stored securely. Finally, all of these are provided in a simple interactive user interface. The hardware components and the operation on data on each block (see Figure 6.3) have separate usage and need some requirements.

3.3 STORAGE AUTOMATION

The energy storage unit is the backbone of the grid. It provides the characteristics of varying demand management to the grid. The batteries have some definite lifespan—maximum charge and discharge cycle. The storage system is the

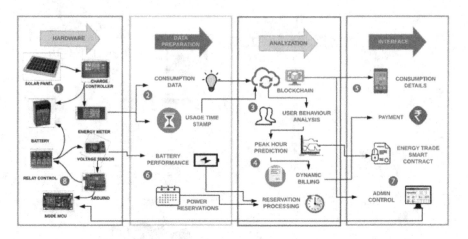

FIGURE 6.3 Energy Distribution Block Diagram.

costliest component of any energy grid. The battery's lifespan plays a significant role in the cost factor; lack of regular maintenance and proper usage may result in less lifespan of cells. In cognitive grid architecture, the system analyzes the operation data for better management. The architecture consists of a battery poll and the relay switch controller for energy flow control of the respective batteries. The voltmeter connected to cells in storage acts as a performance monitor. The current battery energy status and previous charge-discharge cycles are stored for life span prediction and storage management. This storage management unit's primary function is to reduce the wastage of energy in the dead or low-performance cell, so the battery pool is continuously monitored for performance at regular intervals. The voltage sensor measures the battery status using relay control, and the required action to take on bypassing the dead cell is also operated using relay controls.

3.4 Data Analysis

The smart meter fixed in each consumption unit provides the user energy consumption. The data includes information about the amount of energy (in kWh) consumed each day and the amount of energy (in kWh) generated, provided by the meter attached in production. These data are processed in cluster analysis to form a group of users according to their usage behavior. The parameters are the amount of energy consumed by a consumer in the previous month, the amount of energy consumed by a user in the current month, and the duration of energy consumption by that user in peak hour of the grid (see Figure 6.4)—these decide the appropriate group to which a consumer belongs. There is a different billing structure for each of the consumer clusters. The group of consumers who consumes more energy during peak hours is charged higher than the group of consumers who consumes less power during peak hours [14]. Machine learning algorithms predict customer demand. The prediction of the customer usage consumption unit (data) is collected periodically (maybe three times a day); the peak hour utilization is a particular area where the pricing can be optimized based on the user's consumption.

The consumer's average electricity consumption throughout the month and their average use during the peak hours alone are provided to the spectral cluster model.

4 GRID MODULES

4.1 Energy Storage Pool

The batteries in the storage unit are connected with a voltmeter, and it monitors the potential differences in each battery (see Figure 6.5). The original capacity of the battery and the current output is compared for performance calculation. The dead cell in the serial connection can affect all the cells that preceded this dead cell. Hence, regular monitoring identifies the low performing or dead battery—it reduces energy wastages.

Batteries are grouped to form multiple power banks. Battery status is monitored and relayed routes power to and from each bank to maintain the deep discharge cycle as 75%.

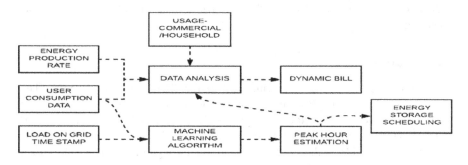

FIGURE 6.4 User Data Analyzation.

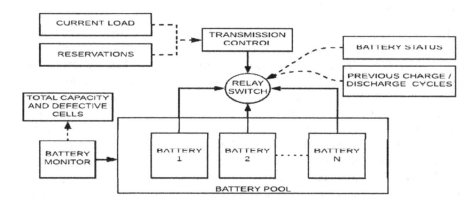

FIGURE 6.5 Energy Storage Management.

4.2 Energy Flow Controller

The analysis process includes the processing of user consumption data provided by the smart meters in the grid for the individual consumers, and the production data provided by the smart meters gives the amount of energy flow from power generators (solar panels and wind turbines) to the grid. Then, a load of the grid is stored for the respective time intervals which are processed by the time series algorithm and the peak hours predicted. The user data analysis is used for dynamic billing and the predicted peak hours are used for scheduling energy storage. The overall billing process includes energy production, consumption, and time stamp data analyzed using clustering algorithms to produce appropriate bills. The relay connected with Arduino controls the energy flow. This control is according to the smart contract deployed in the blockchain.

4.3 Data Storage

The IoT device data is transferred to the IPFS node, then the data is segmented into pieces and stored in a distributive manner over the collective storage location in the IoT network. After that, the content-address is returned to a smart contract. It changes the state of the contract and becomes immutable. Because of that, updates in data are stored as a new file in IPFS, and the address of the latest version gets stored again.

4.4 Communication

The smart meter transmits data wirelessly to the WIFI module ESP8266. The module has a built-in UART (serial) connection. It is used for both sending and receiving data using the UART protocol. The communication follows the TCP/IP protocol.—the communicating module connected with the Arduino board. The battery voltage read from the storage is transferred through this module and the data is forwarded to the DB by the microcontroller.

5 EXPERIMENTAL SETUP

After the setup of the grid, the initial load on the grid entered—the initial load calculated from the usage capacity that is requested by the consumers in the grid. Then, the power generators are operated to serve the initial load by storing the required amount of energy in the storage unit. Before the energy is stored in the battery, the load and production rate of energy in the grid is analyzed. According to that, the battery charging and discharging commands are issued to the storage unit, and then the energy is transferred to consumers in the grid. The smart meter readings are assigned through WIFI communication and power reservation requests from the consumers' smart contracts between nodes in the grid are stored in a blockchain. These data are then retrieved for analysis purposes. According to the data stored, consumers are provided with billing in the user interface application; based on payment details, the relay switch between consumer and power line is

State-of-the-Art Energy Grid 103

controlled for energy flow. The total energy and data flow throughout the system detect the faults and malfunctions quickly. The Voltage sensor reads the voltage at regular intervals and updates the system for low performing battery cells and solar panels. The Relay control eliminates the connection of low performing batteries. The Energy Meter reads the user consumption data and transmits it to the server, and the summation of the meter data of consumption and production provides the live demand variation. The data generated by the IoT devices and consumer data get stored in blockchain architecture, which hashes values with key and then sequentially link blocks with those hash values. The energy trade contract with the prosumers is coded into a smart contract—it automates payment and penalty calculations. Consumers can reserve their additional requirements. The Load data provided by the energy meters with time stamp analyzed with the Machine Learning algorithm to derive the pattern in the consumption from that peak hours predicted. The Users' consumption data further explained to find usage behavior such as their consumption rate and time. So that these analysis results compared with peak hours and production rate to provide appropriate energy prices for the users. The Energy bill payment details live usage monitoring and the power reservation features provided as a web application for the user interface purpose (see Figure 6.6).

6 TECHNIQUES INCORPORATED

6.1 BC Network

The contracts between consumers, prosumers, and the grid management are coded into smart contracts and the SC is compiled to an executable. This code includes automated commands for the IoT devices in the grid, such as relays and smart meters. This contract is a block in the blockchain and is executed for automation. The user consumption data from IoT devices are provided to the blockchain network after being compiled into a smart contract using platforms like Ethereum virtual machine.

6.2 IoT Data Transmission

The communication aspect is classified into two types based on their range. For short-range communication, IoT components such as (Node MCU ESP8266) is used. For long-range communication, we have the plan to use CRN (Cognitive Radio Networks) so that the efficiency of transmission of devices can be improved.

6.3 Smart Contracts

Blockchain architecture is a distributed decentralized network. The primary purpose of the decentralized concept is security and trustless. Keccak-256, the cryptographic hash function, is used to create a digital signature for each unique block in Ethereum Blockchain (EBC) [15]. The smart meter data and the grid load/demand data is hashed with the cryptographic hash function, and this hashed output is stored as a block—this initial block is known as genesis block [16]. The next consecutive

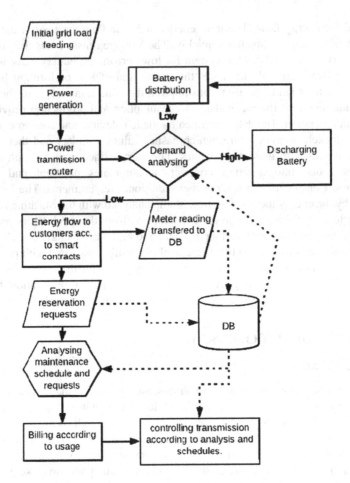

FIGURE 6.6 The Workflow of Data Storage and Power Retrieval.

data storage goes through the same operation, and the addition to the current block hash output apart from the genesis block uses the hash of the previous blocks to create a chain structure [17].

The grid load is stored—along with its timestamps—as a block in the EBC. The trade contract between two parties in the grids is programmed using an object-oriented programming language called Solidity. This contract contains instructions that should be carried out in certain conditions. A private Ethereum blockchain node is installed, and the consumers are provided with an account on that platform. The programmed contract is then submitted to that blockchain node (see Figure 6.7). Then, the contract is compiled into byte code. After that, the compiled byte code is deployed in that blockchain instant, and any tampering with the contract cannot be reflected in that running instant. It ensures a trustworthy deal without a mediator between unknown parties. It also provides automation in the control of IoT devices in the grid [13] and encryption of the data in the blockchain with a hash function.

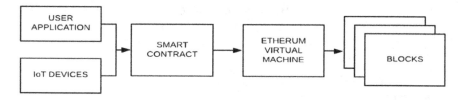

FIGURE 6.7   Data Manipulation.

6.4 DISTRIBUTED STORAGE STRUCTURE

The Interplanetary File System provides a distributed storage structure; the IPFS node is served along with the application. Data from IoT devices pass to that node, and then the data is fragmented for distributive storage in the deployed network. Each node in the network can access and host the contents. After storing the data, the IPFS node provides a cryptographic hash to retrieve that data—this unique fingerprint made of the storage location and data itself. Hence, the hash act as an address. IPFS node removes duplication across the network, which supports the effectiveness of application in data handling. Interplanetary Name System hosted with the IPFS enables manual access for the data that is in human-readable names. IPFS delivers the peer-to-peer approach for data transmission, which results in effective communication in the massive volume of data.

6.5 COGNITIVE CHARACTER

Cognitive skills and knowledge are the ability to acquire information, often the kind of expertise that is easily tested—cognition distinguished from social, emotional, and creative development and ability. Cognitive science is a growing field of study that deals with human perception, thinking, and learning. The reason for naming the system as Cognitive Grid is that it has a set of skill sets that align with cognitive behavior, such as intellectual functions, judgment, computation, evaluation, reasoning, decision making, and problem-solving. Also, cognitive technologies can be used to support communication in the long-range [18]. The appropriate channel selection reduces packet loss and interference in connection.

7 CHALLENGES AND ISSUES

In the smart energy grid, there is a huge volume of data to handle, which requires more storage space for effective operation in large scale grids. The IoT data transmission may result in high communication traffic in the system. While implementing the latest technology, the market may hesitate to adopt the new system. The degradation of fossil fuels and carbon-footprint of energy generation has moved to the phase where production costs us a high price not only economically but also geographically—on top of life risk factors. The aim is to contribute to the

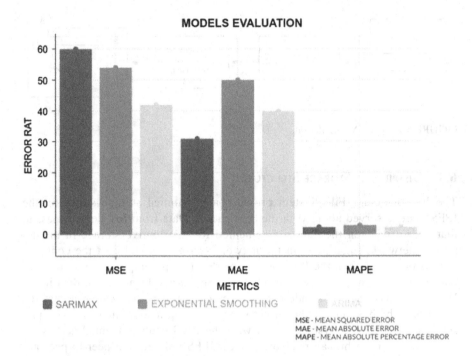

FIGURE 6.8 Models Evaluation Result.

technologies and concepts that are working forward to reduce the factors that drag the replacement of non-renewable energy with renewable energy. Commercial implementation of a self-restrained grid reduces hazardous fossil fuel burning. During natural calamities, distributed architecture meets the energy requirements of the hospitals for communication purposes. Moving to renewable resources reduces carbon-footprint and greenhouse gases.

8 RESULTS

8.1 SARIMAX Load Forecasting

The Seasonal Autoregressive Integrated Moving Average with external variables (SARIMAX) forecasted load more accurately than the Autoregressive Integrated Moving Average Model (ARIMA) and exponential smoothing (see Figure 6.8) (which is generally used for peak-hour predictions). The models mentioned above are evaluated with Mean Squared Error (MSE), Mean Absolute Error (MAE), and Mean Absolute Percentage Error (MAPE) metrics [19] (see Figure 6.9).

Among the three-evaluation metrics, results in the SARIMAX model has a low error percentage in two areas (see Table 6.1). The prediction result also shows that the SARIMAX model forecasts better than ARIMA and exponential smoothing as the number of training sets increased.

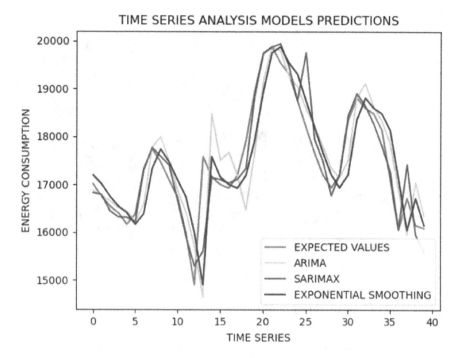

FIGURE 6.9 Predicted and Expected Result Comparison.

TABLE 6.1
Average Error over a Different Number of Datasets (100–1,000)

Time Series Analysis Models	Avg. MSE	Avg. MAE	Avg. MAPE
ARIMA	417651.64	398.80	2.315
SARIMAX	604216.77	309.23	1.831
Exponential Smoothing	539743.28	499.79	2.885
BEST	ARIMA	SARIMAX	SARIMAX

8.2 Spectral Clustering

The spectral clustering of users according to their peak hour consumption and contribution resulted in certain groups that provide appropriate bill structures. The spectral cluster model is compared with popular clustering algorithms such as K-means and Min batch K-means and performed better than these. The K-means algorithm is a basic and generally used algorithm for clustering purposes [20], but handling several attributes and parameters requires a more functional approach. The data points are accurately grouped according to the distance from one another. Performance is analyzed with the silhouette score metrics (see Figure 6.10) used to compare the clusters created by the three models mentioned above

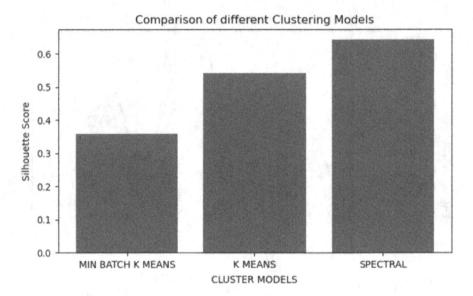

FIGURE 6.10 Cluster Accuracy of Models.

TABLE 6.2
Silhouette Scores for Cluster Accuracy

Clustering Models	Silhouette Score for 50 Data Points	Silhouette Score for 100 Data Points	Silhouette Score for 150 Data Points
Min-batch K means	0.566	0.548	0.548
K means	0.561	0.631	0.464
Spectral	0.579	0.677	0.703

(see Table 6.2). The silhouette score ranges from 0-1; the closer to 1 the value, the more accurate the cluster is.

8.3 Distributed Storage

The InterPlanetary File System significantly increases data availability and accessing speed. It highly supports systems with frequent data transmissions. Along with blockchain, this content-based storage provides data privacy as well as security.

8.4 P2P Energy Trade with Reservations

The prosumers concept is developing gradually, and the P2P structure supports its development and reduces considerable losses in energy during far distance transmission. The power reservation feature enables grid operators to utilize each resource with minimal loss.

9 CONCLUSION

The investment in renewable energy sources will recover slowly in the future, but adapting renewable energy will have a significant impact on environmental pollution and energy demands. Once the investment recovers, the energy harvested only with the maintenance cost can boost the profit rate. Energy distribution fields adopted the latest technologies in recent years for commercial purposes. Now, there is a situation in need of an entirely eco-friendly energy production. We need an efficient distribution system to reduce the gap between energy consumers and renewable sources. The limitations—not addressed in previous works on grid architecture—are discussed in this paper. Some factors compromise the effective operation of the grid, but that will be refactored in future enhancements.

REFERENCES

1. Al-Turjman, F. and Abujubbeh, M. (2019). IoT-enabled smart grid via SM: An overview. *Future Generation Computer Systems*, 96, 579–590.
2. Rathee, G., Sharma, A., Kumar, R., and Iqbal, R. (2019). A secure communicating things network for industrial IoT using blockchain technology. *Ad Hoc Networks*, 94, 101933.
3. Faheem, M., Shah, S. B. H., Butt, R. A., Raza, B., Anwar, M., Ashraf, M. W., ..., Gungor, V. C. (2018). Smart grid communication and information technologies in the perspective of Industry 4.0: Opportunities and challenges. *Computer Science Review*, 30, 1–30.
4. Noor, S., Yang, W., Guo, M., van Dam, K. H., and Wang, X. (2018). Energy demand side management within micro-grid networks enhanced by blockchain. *Applied Energy*, 228, 1385–1398.
5. Fadel, E., Faheem, M., Gungor, V. C., Nassef, L., Akkari, N. , Malik, M. G. A., ... Akyildiz, I. F. (2017). Spectrum-aware bio-inspired routing in cognitive radio sensor networks for smart grid applications. *Computer Communication*, 101, 106–120.
6. Li, Y., Rahmani, R., Fouassier, N., Stenlund, P., and Ouyanga, K. (2019). A blockchain-based architecture for stable and trustworthy smart grid. *Procedia Computer Science*, 155, 410–416.
7. Mhaisen, N., Fetais, N., and Massoud, A. (2019). Secure smart contract enabled control of battery energy storage system against cyber-attacks. *Alexandria Engineering Journal*, 58, (4), 1291–1300.
8. Tripathy, A.K., Das, T. K., and Chowdhary, C. L. (2020). Monitoring quality of tap water in cities using IoT. In *Emerging Technologies for Agriculture and Environment* (pp. 107–113). Springer, Singapore.
9. Li, Y., Yang, W., He, P., Chen, C., and Wang, X. (2019). Design and management of a distributed hybrid energy system through smart contract and blockchain. *Applied Energy*, 248, 390–405.
10. Foti, M. and Vavalis, M. (2019). Blockchain-based uniform price double auctions for energy markets. *Applied Energy*, 254, 113604.
11. Tripathy, A. K., Mishra, A. K., and Das, T. K. (2017, July). Smart lighting: Intelligent and weather adaptive lighting in street lights using IoT. In *2017 International Conference on Intelligent Computing, Instrumentation and Control Technologies (ICICICT)* (pp. 1236–1239). IEEE.

12. Kochovski, P., Gec, S., Stankowski, V., Bajec, M., and Drobintsev, P. D. (2019). Trust management in a blockchain-based fog computing platform with trustless smart oracles. *Future Generation Computer Systems*, 101, 747–759.
13. Roux, M. and Booysen, M. J. (2017, April). Use of smart grid technology to compare regions and days of the week in household water heating. In *2017 International Conference on the Domestic Use of Energy (DUE)* (pp. 276–283). IEEE.
14. Basumatary, J., Singh, B. P., and Gore, M. M. (2018, January). Demand side management of a university load in smart grid environment. In *Proceedings of the Workshop Program of the 19th International Conference on Distributed Computing and Networking* (pp. 1–6).
15. Jindal, A., Aujla, G. S., and Kumar, N. (2019). SURVIVOR: A blockchain-based edge-as-a-service framework for secure energy trading in SDN-enabled vehicle-to-grid environment. *Computer Networks*, 153, 36–48.
16. Alam, S., Aqdas, N., Qureshi, I. M., Ghauri, S. A., and Sarfraz, M. (2019). Joint power and channel allocation scheme for IEEE 802.11af based smart grid communication network. *Future Generation of Computer Systems*, 95, 694–712.
17. Pieroni, A., Scarpato, N., Di Nunzio, L., Fallucchi, F., and Raso, M. (2018). Smarter city: Smart energy grid based on blockchain technology. *International Journal on Advanced Science Engineering Information Technology*, 8(1), 298–306.
18. Khan, M. W., Zeeshan, M., Farid, A., and Usman, M. (2020) QoS-aware traffic scheduling framework in cognitive radio-based smart grids using multi-objective optimization of latency and throughput. *Ad Hoc Networks*, 97, 102020.
19. Panda, S. K., Bhoi, S. K., and Singh, M. (2020). A collaborative filtering recommendation algorithm based on normalization approach. *Journal of Ambient Intelligence and Humanized Computing*, 1–23.
20. RM, S. P., Maddikunta, P. K. R., Parimala, M., Koppu, S., Reddy, T., Chowdhary, C. L., and Alazab, M. (2020). An effective feature engineering for DNN using hybrid PCA-GWO for intrusion detection in IoMT architecture. *Computer Communications*.

7 Optimized Channel Selection Scheme Using Cognitive Radio Controller for Health Monitoring and Post-Disaster Management Applications

R. Rajaguru and K. Vimaladevi[2]

Department of Computer Science and Engineering, Sethu Institute of Technology, Virudhunagar, Tamil Nadu
School of Computer Science and Engineering at Vellore Institute of Technology, Vellore

1 INTRODUCTION

Due to the increasing usage of wireless services globally, network availability has been facing resource scarcity and heavy traffic occurs for communication. Consequently, it leads to two significant issues — miscommunication and delay in the network. The remote monitoring system has been involved in different applications such as health monitoring, agriculture, industrial automation, defense, telecommunication, and post-disaster management. Due to the increase of wireless services and IoT communication, the availability of channels has decreased drastically, and it leads to the non-availability of channels for an emergency. Nowadays, with the help of sensors and intelligent devices, the device terminal offers a connection between different machines and provides communication, and it is known as Machine-to-Machine communication (M2M) [1] and [2]. The critical reason for the usage of M2M communication is that it utilizes less energy, increases system capacity, and maximizes the spectrum efficiency. Currently, M2M communication is performed using ISM bands, which are known as unlicensed bands (i.e. the bands possess no fee for licensing). For example, Wi-Fi is meant for private usage. Because of the increasing utilization of M2M communication, the density of the network is maximized, and it leads to traffic as well as collisions in the wireless networks.

Based on Cooper's law, the usage of data transmission and voice over services has increased twice every 30 months. Because of the exponential usage of handheld devices, effective spectrum allocation, and utilization have become challenging tasks in wireless services and communications. As a result, the Next Generation Network will offer solutions to this exponential growth of wireless device users and provide a good quality of service to these users. It is essential to deliver traffic-free communication and maximize the utilization of available radio resources to carry out all kinds of wireless services. In recent years, the world has been facing many natural calamities owing to natural and human-made disasters such as floods, earthquakes, tsunami, gas leaks, chemical explosions, etc. As a result, the wireless infrastructures are profoundly affected and, in turn, the networks are pulled down, leading to improper communication. The post-disaster management deals with saving the lives of disaster victims. In health monitoring applications, Wireless Body Area Network (WBAN) monitors the patient's health and transfers the information to the data pool without influencing the patient's daily routine activities. Increasing such kinds of wireless services and wireless technologies will cause a higher demand for radio resources. The Federal Communication Commission (FCC) has announced that most of the available licensed radio resources are under-utilized by the owner or primary users. In traditional wireless networks, certain spectrum bands are utilized maximum by the user in specific geo-locations and time frames; some spectrum bands are not used. This approach leads to high traffic and unbalanced utilization of most of the valuable radio resources in a dynamic environment. Besides, maximum collision occurrence is probable in unlicensed bands due to the increased number of users using the internet and wireless services for their applications.

In a multi-cloud aspect, to improve the quality of service in the different condition, novel uncertainty-based QoS algorithms were proposed by the authors Sanjaya K. Panda and Prasanta K. Jana. This algorithm mainly focuses on task scheduling with uncertainty aspects, and the authors proved that the running time for this algorithm is better compared to those of SBTS, CMMS, and CMMN [3]. The regulatory authorities' policies employ static spectrum allocation methods and assign a new spectrum band for offering new kinds of services to the users. These approaches lead to reduced utilization of available spectrum bands. To improve the usage of spectrum, the policies of regulatory authority allow the usage of licensed bands by the non-licensed user, subject to the condition of without causing interference to the licensed users. To provide solutions to the spectrum scarcity problem and to efficiently utilize the spectrum and radio resource management, Mitola has introduced the concept of cognitive radio (CR), which can learn the current radio frequency environment and its surroundings. Then, it autonomously changes the operational parameters based on the observation to access the radio (frequencies) that are idle. In general, the CR consists of different components such as radio, sensors, knowledge database, and two various engines for learning and reasoning. The main feature of CR is that it can reconfigure transmission parameters such as modulation techniques, transmission power, and RF. There are three distinct stages, such as observe, learn and reason, and act in the cognitive cycle, as shown in Figure 7.1.

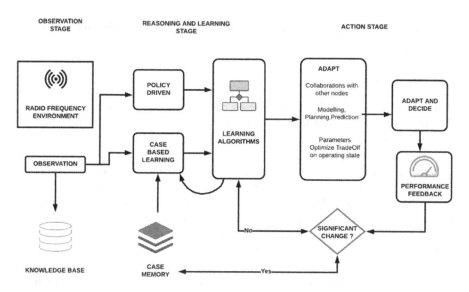

FIGURE 7.1 Operations in the Cognitive Cycle.

1.1 COGNITIVE RADIO NETWORK

Recent advancement in the cognitive radio network (CRN) is that it offers an un-utilized frequency band to the cognitive user with prior permission of the spectrum owner. The spectrum owner calls it the primary user who has full privilege to use the allocated spectrum without any interference caused by any devices or services. The cognitive user/secondary user can use an un-utilized frequency band concerning the period and geo-location with permission without any interference to the primary user. The underutilized frequency channels need to be used effectively to maximize the usage and management of radio resources on the internet, web, and wireless services. Hence, CRN is a promising approach to meet the spectrum demand and radio resource management for the forthcoming next-generation networks.

A spectrum hole or unused frequency band or idle channel is defined as primary users who do not utilize the allocated spectrum for any services in a certain period in a specific geo-location. To improve the efficiency of the radio resources and to effectively use the spectrum, the identified idle channel can be assigned to the unlicensed user to make adequate dynamic spectrum access (DSA) for CR communications.

In the recent development of CRN, it is applied in different applications for the effective utilization of the available spectrum for health monitoring applications, military purpose, vehicular networks, Delay Tolerant Networks (DTNs), etc. CR communication also enables and restores the connectivity of the network when the built-in infrastructure is destroyed or disabled by natural disasters in disaster-response, emergency, and public safety networks by using the existing spectrum without requiring any support. All these kinds of applications need effective DSA with efficient routing protocols to make useful data transmission without any delay and packet loss. To improve the DSA, good channel selection and decision techniques with suitable routing processes are required to utilize the spectrum hole [2] efficiently.

1.2 Dynamic Spectrum Access

To improve the DSA, CR communication has adopted the following sequence of actions: sensing the idle channel (spectrum sensing), accessing that identified idle channel (spectrum access), and managing the channel used for the cognitive users (spectrum mobility functions).

Sensing: In CRN, the primary function is to identify the free nodes belonging to different frequency ranges surrounding the mobile users' radio environment, check the availability of the node, and find out whether the primary user is using it or not. The sensing information also specifies the primary user's usage details, which supports the identification of the free spectrum hole.

Identification: It denotes the identification of spectrum holes in the sensing process using energy detection techniques. Then, it analyzes the behavior of the identified spectrum hole based on the primary user's usage and learns its activities. It also finds out the information of similar spectrum holes.

Access: CR user must decide which spectrum hole has to be used from the collected spectrum holes identified. Based on the decision, the CR user must reconfigure the parameter to access the identified spectrum hole.

Mobility: While using the spectrum holes, suppose when the primary user must access the licensed hole, the CR user must vacate the used spectrum hole immediately and switch to the next available spectrum hole from the identified collections.

In a wireless communication network, the available spectrum bands have limited resources. The main objective of the Next Generation Networks (NGN) will focus on how to utilize spectrum efficiently. The NGN [4] and [5] will need to support the following aspects, such as increasing spectral efficiency, reducing energy utilization, minimizing latency to achieve maximum data rate, and improving the underutilization of spectrum through DSA techniques.

The overall functionality of DSA in CR communication is shown in Figure 7.2. As per the conceptual view of Joseph Mitola to implement the CR technologies, a software platform named Software Defined Radio (SDR) is built. The platform helps in programming and reconfiguring the radios, built on a digital signal processor or a general-purpose microprocessor.

Based on the sharing aspect, the DSA is classified into three models: common-use, shared-use, and exclusive-use.

Common-use model: The entire frequency band is open for all types of users.

Shared-use model: In the absence of the primary user's allocation to channels, the unlicensed user can use those channels with a condition to not interfere with primary users.

Exclusive-use models: The frequency bands owned by the licensed owners will allow the unlicensed user to use their frequency channel even in their presence for a particular period and geographical locations. Using spectrum underlay and spectrum overlay approaches, the cognitive users can use both the

Optimized Channel Selection Scheme

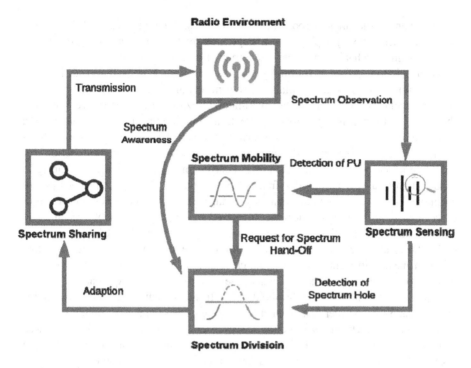

FIGURE 7.2 Flow of Dynamic Spectrum Access Techniques.

licensed and unlicensed bands using dynamic spectrum access techniques, as illustrated in Figure 7.2.

2 PROBLEM IDENTIFICATION

The static spectrum allocation method leads to the underutilization of the spectrum owned by the current user, which leads to the spectrum scarcity problem. To overcome these issues, the next-generation network dynamically allocates the spectrum with the help of CRN. The CRN that allows the cognitive user to explore the spectrum that is not used by the spectrum owner can devise an approach to access those spectra dynamically. For the effective utilization of the spectrum by the cognitive user, channel selection decision plays a vital role. To improve performance, channel selection can be made by the following sequence of steps: having known the characteristics of the channel, channel selection and reconfiguration of parameters to access the channel are important for the spectrum decision in CRN, as suggested by the authors M. T. Masonta et al. [6]. To achieve effective DSA, the channel selection approach for cognitive users plays a vital role. Allocating the best idle channel will improve resource utilization and CR communication. To provide the best channel selection, the authors Gulnur Selda Uyanik and Sema Oktug [7] proposed a primary user temporal-activity-estimation-aided spectrum assignment scheme. In their scheme, the data transmission for cognitive users will be directly correlated with temporal usage of that particular primary user channel for the channel allocation purpose in a distributed scheduling manner. This channel assignment scheme was deployed in a single and multi-channel environment. An effective routing mechanism provides uninterrupted data transmission.

To provide an effective routing mechanism, Rakesh Kumar Jha et al. [8] proposed a model to utilize the available spectrum recourses for dynamic access and it is a key technique to achieve the maximum utilization of an effective spectrum sharing. Spectrum sensing, network selection, and channel identification plays a vital role in improving the spectrum-sharing approach. The authors proposed a four-layered approach: spectrum sensing, network selection, channel allocation for cognitive users, power optimization and security for effective spectrum sharing.

G. Singh et al. [9] proposed a framework to improve the throughput analysis of cognitive users in CRN using probability channel selection techniques. In their proposed approach, the performance was improved by selecting a suitable channel and establishing communication over the channel as a joint operation in CRN. Usually, in CRN, the channel selection process for the cognitive users is done using a random selection method to identify whether the channel is idle but, in their work, they proposed two different approaches for channel selection in CRN. In Approach 1, channel selection is made based on the threshold value. If the particular channel's threshold is greater than 50%, consider that channel for sensing purpose in a usual random manner. In approach 2, the channel selection is based on predicting the high probability of an idle state channel. Using these approaches, the authors proved that the throughput of CRN is improved when compared to a random channel selection approach.

Sadip Midya et al. [10] differentiated the cognitive users into two different categories, such as Real-Time User (RT) and Non-Real Time User (NRT). RT users have a higher privilege of data transmission as compared to NRT users. For RT user's data transmission, they use Voice over Internet Protocol (VoIP) along with some gaps. In the same data channel, NRT users make data transmission using the silent gaps. To implement these aspects, the authors used compatibility function in binary mode and deployed in the gaming model, namely the auction game model.

Prioritized Secondary User Access Control approach has been adopted to reduce the performance of the spectrum sensing process if jamming signal was injected in data channel by the primary users or unauthorized cognitive users to reduce the accuracy of sensing results in CRN. To improve the quality of service of cognitive users in CRN, an integrated approach has been adopted to select channels, which has been considered for channel decisions also. By this approach, the channel allocation process not only depends on the availability of the channel but also considers other aspects such as using the history of the channel for the primary users, while making the decision. It is proposed by the authors L. Jayakumar and S. Janakiraman [11]. In this approach, the authors proposed a scheme for selecting a suitable channel by integrating the EFAHP-SAW and EFAHP-MOORA approach to reduce the number of switchovers by minimizing the time for computational analysis.

In general, the reduction of the bit error rate, utilization of energy, and avoidance of interference to primary users in CRN was based on the cooperative spectrum sensing approach. In the CSS approach, the selection of cognitive users was an important task to minimize these values and to improve the throughput and spectrum usages. Optimized channel selection for cognitive users is an interesting research topic. Based on the SNR values, the authors Sudhamani Chilakala and M. Satya Sai Ram [12] proposed the AND and OR fusion rules. If the SNR values are minimum, apply the OR fusion rules, otherwise, apply the AND fusion rule for the maximum SNR values. Through these fusion rules, the authors proved that there is an increase in the usage of spectral efficiency by their experimental setup.

To achieve collision-free channel allocation methods for D2D communication for dynamic environmental conditions, the authors Haitao Zhao et al. [2] proposed a sender-jump blind rendezvous channel allocation approach. In their approach, the sender blindly jumps and identifies their received receiver node assigned for data transmission with the help of a handshake procedure. Here, the sender's node jumps to different nodes to reach the receiver part.

The intelligent/cognitive radio was used in next-generation networks to manage the needs of wireless services and provide a better solution to the spectrum scarcity problems. CR plays a vital role in avoiding the scarcity problem by providing the optimal radio resource allocation to the unlicensed users without disturbing the licensed users. But increasing the number of wireless users/services will create a problem in radio resource distribution and channel allocation to the unlicensed users in CRN due to the primary traffic in the network. The cognitive user uses radio resources in the absence of primary users, and in the sudden arrival of the primary user, the cognitive user will vacate the radio resources. Due to this dynamic nature of the cognitive user, it needs an efficient spectrum access mechanism. DSA, along with an intelligent spectrum access coordination, will optimize this spectrum with minimum delay for the cognitive user [10]. So, designing an effective dynamic spectrum access mechanism to improve the accuracy and system throughput of the CRN is a necessary task.

To optimize the decision results on the Internet of Things-based Fire Detection System and Fall Detection System based on the Internet of Things, the authors Sourav Kumar Bhoi et al. proposed a new K-Nearest Neighbors (K-NN) and decision tree in machine learning algorithms. Using this algorithm, the author improved the accuracy in detection aspects. The simulation results also proved the accuracy of 93.15% and 89.25%, respectively, by K-NN and decision tree in the machine learning algorithm [13] and [14].

To effectively utilize the spectrum hole in CR communication [15,16] and [17] identify the idle channel plays a significant role in improving the routing mechanism's performance and stability. The idle channel selection technique identifies the channels with high availability, high connectivity with neighbors, and low primary user activity, and then the routing will be very efficient and stable. Further, it exists for a longer time. Therefore, channel selection should be properly investigated for routing in CRN and, in turn, routing will be performed effectively and efficiently [18,19].

The main objectives of the proposed DSA are

- To provide the most suitable and optimized channel selection for the spectrum mobility to maximize the performance of DSA.
- To provide an efficient and effective dynamic spectrum access approach for cognitive users.
- To provide more secure data transmission over heterogeneous networks.

- To provide an assistant system for medical professionals to monitor paralyzed patients through the health monitoring system.
- To provide an assistant system for the post-disaster management team to transfer the information without any delay.

3 SYSTEM MODEL

This proposed DSA approach mainly focuses on channel identification, channel assignment, and channel mobility as shown in Figure 7.3; the spectrum mobility is done with proper handoff procedure adopted for the cognitive users and the SH based on scheduler and link routing algorithms. The proposed approach predicts the idle channel using the energy detection method in CSS approach for the decentralized CRNs after identifying the idle channel based on channel parameters such as channel identification, channel capacity, channel switching delay, channel interference, channel holding time, error rate, subscriber location, and path loss. This will show the list of idle channels that are used by cognitive users. From the list of idle channels, arrange the idle channel from the highest probability to the lowest probability to allocate to the cognitive users. Suppose the predicted channel is in an inactive state, then select the next random channel for sensing technique.

After identifying the possible list of idle channels, the system adopts the channel optimization techniques that produce the optimal channel selection for the cognitive users. The optimized channel list correlates the channel parameter with user parameters for better channel assignment to the cognitive users. If the sudden arrival of primary users or a reduced the quality of the selected channel, then the system must adopt the channel mobility task that helps the cognitive users identify the next target channel for CR communication.

3.1 CHANNEL IDENTIFICATION

The efficiency of the DSA in CRN always depends on identifying the idle channel for cognitive users. Sometimes, improper channel identification reduces the performance of DSA; identification of the idle channel is a vital one. The proposed approach adopts the energy detection method with a cooperative spectrum sensing approach. The energy detection method is a more suitable approach because it does not need any prior information about the channel. Based on the threshold value, it is predicted whether the primary users use the channel or not. If the threshold value is minimum, then it is decided that the respective channel is idle. The binary value represents the channel — if it is 0, then that channel is idle; if it is 1, then that channel is used by the user. Figure 7.4 illustrates the channel identification approach using the energy detection method.

3.2 CHANNEL ASSIGNMENT

To improve the performance of cognitive users, channel access plays a vital role in the proposed system. The channel allocation is done based on the list of available idle channels from the channel identification process. Then, the possible channels

Optimized Channel Selection Scheme

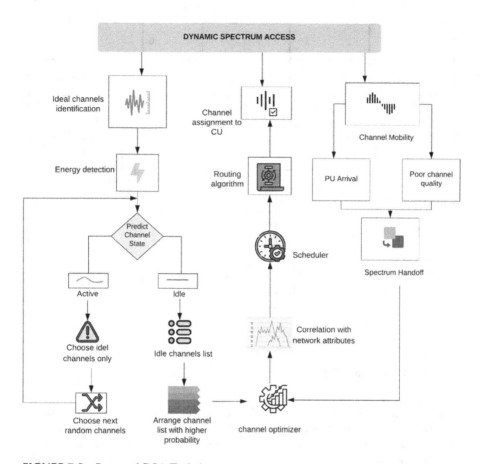

FIGURE 7.3 Proposed DSA Technique.

are listed from the highest to lowest possibility to accommodate the cognitive users. Here, the optimization technique is carried out between the channel quality parameters and cognitive users, which require the idle channel to carry out the CR communications. The optimization is done with the available list of channels with the cognitive users requiring the idle channel. From this, the optimal channel can be allocated to cognitive users. By assigning the optimal channel to cognitive users, the unnecessary switching time is reduced, which can improve the execution time for CR communication.

3.3 Channel Mobility

To improve the efficiency of radio communication using CR, spectrum mobility in CRN is an important task that can occupy and use the unused channel or frequency band of the primary user to the cognitive user by using the opportunistic manner in a dynamic environment. CR communication should assure that there is no cause of the harmful interference to the spectrum owner at any period and geo-location, because of

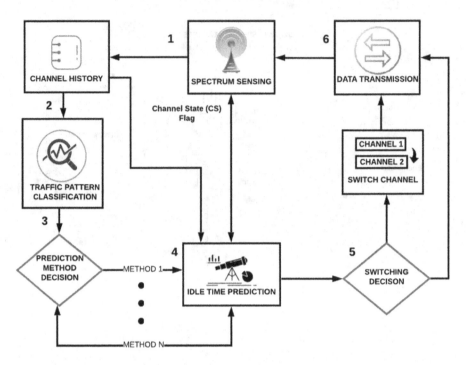

FIGURE 7.4 Spectrum Hole Detection.

the primary user's activities having the highest privilege. The sudden arrival of the primary users leads cognitive users to vacate and handover the channel. To improve CR communication, the cognitive users will find the next new target channel to continue. This process is known as spectrum handoff (SH). The researchers developed many SH algorithms for CRNs to provide a solution to the cognitive users for the arrival of primary users. To complete the unfinished CR communication, SH algorithms suggest new target channels and some interrupt mechanisms for switching the CR data to the new target channel from the transmission buffer for the completion of on-going communication for the cognitive users.

4 SCHEDULING AND ROUTING PROCESS

The performance of the cognitive users is focusing on effective CR communication; CR communication deals with DSA consisting of e-channel identification, selection, assignment, and handoff. If there are more cognitive users, then the system must allocate the idle channels based on the scheduling process and effective linking and routing mechanism. In our system, scheduling decisions are made during all interval time slots, even during the dynamic networking environmental conditions. The scheduler S_{if} values denote the maximum numbers of packets per time slot that can be sent by SUi using frequency f during the entire scheduling period are found. Reliable communication of the secondary users with the Cognitive Relay Station (CRS) and Cognitive Base Station (CBS) is ensured.

4.1 Transmission Power

Here, the transmission power P_t^{ift} is considered the maximum permissible to assertive power for SUi using frequency in time slot t.

$$P_t^{ift} = \min(k \in_{CRS}^{ft}) \frac{P_t^{fk}}{\left(\frac{\lambda}{4\pi d_{ikt}} \times |h_{ikt}|\right)^2} + \min(k \in_{CBS}^{ft}) \frac{P_t^{fk}}{\left(\frac{\lambda}{4\pi d_{ikt}} \times |h_{ikt}|\right)^2}$$

where h_{ikt} is the fading coefficient of the channel $|\phi_{CBS}^{ft}| > 1$ and $|\phi_{CRS}^{ft}| > 1$ because the primary user can use two same frequencies simultaneously.

$(\frac{\lambda}{4\pi d_{ikt}} \times |h_{ikt}|)^2 \rightarrow$ path loss of the channel.

The reliable communication between the secondary user and CRB are

$$R_{ift} = \left[\ln\left(1 + \left(\frac{\left|P_t^{ift} \times \frac{\lambda t}{4\pi d_{it}^{CRS}} \times h_{it}^{CRS}\right|^2}{\varepsilon}\right) + \left(\frac{\left|P_t^{ift} \times \frac{\lambda t}{4\pi d_{it}^{CBS}} \times h_{it}^{CBS}\right|^2}{\varepsilon}\right)\right)\right]$$

R_{ift} – Maximum number of packets that can be sent by the secondary user
d_{it}^{CRS} – Distance between secondary user and CRS
d_{it}^{CBS} – Distance between the secondary user and CBS

4.2 Minimizing Overhead Problem (MOP)

Transmission overhead using different frequency

$$ETX = \frac{\text{No. of packets delivered}_{CRS}}{\text{Total no. of packets}} + \frac{\text{No. of packets undelivered}_{CBS}}{\text{Total no. of packets}}$$

$$= \frac{P_{CRS}}{P_{total}} + \frac{P_{CBS}}{P_{total}}$$

$$ETX = \frac{\delta}{\delta x} \frac{P_{CRS} \text{ to } P_{CBS}}{P_{total}}$$

$$\text{Reliable link quality} = \frac{\delta}{\delta x} \frac{P_{CRS} \text{ to } P_{CBS} \text{ retransmission}}{P_{total}}$$

Retransmission occurred in t seconds.

4.3 Overhead

$$\text{Complexity} = \frac{\text{No. of retransmission nodes}}{\text{System channel capacity}}$$

$$= \frac{RP_{CRS} + RP_{CBS}}{SU_{channel} + P_{total}}$$

$$C = \text{Users density} \sum_{i=1}^{n} \frac{P_t^{ift}|4\pi d_{it}^{CRS} \times h_{it}^{CRS}|^2}{\varepsilon}$$

Link Quality Maintenance Scheduler (LQMS)
Extensible Transmission Rate (ETX) → QoS parameter

Algorithm 1   TLQ MAX | £ TOD MIN for Routing and Linking

Require: N, F, L, a_i ∀I ∈ N
  Ensure: T_x, R_{if} values ∀I ∈ N, ∀f ∈ N
  Step 1. Build an edge-weighted link quality (Multihop) groups G = (M,L,N,E) **as follows.**
  Step 2. For each I ∈ N add a vertex m_i to M.
  Step 3. For each I ∈ N add a link quality vertex l_i to L.
  Step 4. For each f ∈ F add a vertex n_i to N.
  Step 5. For each set of vertices m_i ∈ M, l_i ∈ L, and n_i ∈ N add the edge $\{m_i, l_i, n_i\}$ to E with weight W_{itf}.
  Step 6. Define the following function L which associates function interval of natural numbers with each vertex index.
  Step 7. $L(m_i) = [1, a_i.T] \forall_i \in N$
  Step 8. $L(n_f) = [0, T] \forall\, f \in F$
  Step 9. $L(l_i) = [0, l_i.T] \forall\, l_i \in L$
  Step 10. Find maximum weighted L – factor M of G.
  Step 11. For all i ∈ N and f ∈ F.
  Step 12. For all i ∈ N, f ∈ F, l ∈ L
  Step 13. let X_{it} be equal to the number of edges.... vertices m_i, n_f, l_i in the reliable factor M.

The algorithm deals with the linking and routing mechanism, and that is done by building the graphs with vertexes and links. Here, linking is done based on the sum of the weights between the edges. By these weighted edges, the linking is done dynamically and made for effective CR communications.

TABLE 7.1
Simulation Parameters and Values

Simulation Parameters	Range/Parameter Values
Network setup	
The radius of the network setup	150 m
Coverage radius for the licensed users	75 m
Coverage radius for the CR user	50 m
Propagation model	Line of sight
Number of licensed users	12
Number of CR user	10-100
Other Parameters	
Frequency	2400 MHz
Number of channels	2-20
Packet rate	Random distribution
Probability of sensing idle channel energy	180 µJ/s
SINR	1-14 dB
Size of the packets	45 byte
Transmission power	10 MW
Modulation	OFDM
Time slot length	90 ms
Path loss	2

5 EXPERIMENTAL SETUP AND RESULTS

The simulation environment for the proposed system includes sensors, primary users, cognitive users, control nodes, and base station for the primary users. Since primary users have the highest priority for communication, cognitive users and other nodes can only use the channel to communicate during the absence of primary users. The cognitive users will help as a relay station to sensor nodes when it is in an idle state. The sensor node will also act as a relay state for cognitive users while any information is not sensed.

In Table 7.1, simulation parameters for the proposed system were shown. The number of primary users is 12, and the number of cognitive users varies from 10 to 100.

To identify the spectrum hole, the cognitive user uses cooperative spectrum sensing with the energy detection method. Based on spectrum sensing, the threshold value identifies whether the user is active or not, by which the cognitive user identifies the available spectrum hole. Figure 7.5 illustrates the cooperative spectrum sensing techniques with the probability of detection (Pd) and the probability of misdetection (Pmd).

The time taken to identify the idle channel plays a vital role in CR communication. The efficiency of DSA depends on the time taken to identify the idle channel to improve communication aspects. Figure 7.6 illustrates the time taken to identify the idle channel for varying cognitive users in a dynamic environment. This graph illustrates that when the number of cognitive users increases, the time taken to identify the spectrum hole also increases.

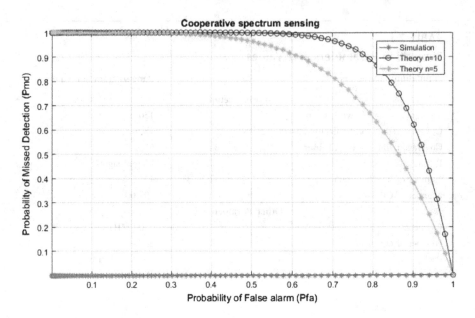

FIGURE 7.5 Cooperative Spectrum Sensing.

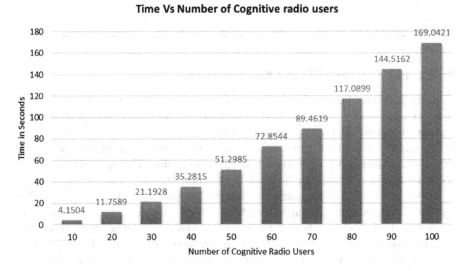

FIGURE 7.6 Execution Time to Identify the Idle Channel.

6 APPLICATION OF DSA IN HEALTH MONITORING SYSTEM

Wireless communication plays a dominant role in networking technologies. Wireless data transmission has been used in various fields for digital norms and efficient results. In a healthcare application, there has been an evolution of WSN. This WSN has some limitations such as the number of access points, data loss, etc., and these constraints can be solved through WBAN. The WBAN comprise different connected sensors imparted in the human body—either inside or outside—to monitor the health record information indirectly.

The existing WBAN system has been comprised of three component-based layers, such as Intra-BAN, Inter-BAN, and beyond-BAN layers. The intra-BAN layer consists of biosensors, a central node, and a cognitive controller — they act as a data extraction layer. The CR device in this layer communicates with the access points or base station, or satellites for node communication. The beyond-BAN layer acts as a bridge between the two layers, and it transfers information to the medical personnel and database. This system selects the best network for communication, but it does not select the optimal network without data loss and network traffic. The best network must have basic characteristics such as faster communication, no data corruption, efficient use of spectrum, etc.

In the present work, three-tier architecture with some additional functions other than just transferring information to the destination has been proposed, and it is shown in Figure 7.7.

6.1 DATA EXTRACTION USING CRC (TIER 1)

The data extraction layer comprises three sensors—temperature sensor, blood pressure sensor, and pulse rate sensor—configured with the central node to act as a controller, as shown in Figure 7.8. These sensors produce analog signals as output. The signals of the individual sensors are transmitted to the central node of the WBAN controller. In the proposed work, Raspberry pi3 has been used as a WBAN controller.

Figure 7.8 shows the sensors connected over the patient's body, transmitting the collected information to the Raspberry PI board, which acts as a WBAN controller over wireless medium using Wi-Fi communication. The data acquired by the sensors are

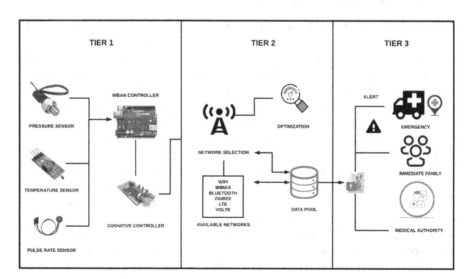

FIGURE 7.7 Three-Tier Architecture for Health Monitoring Application.

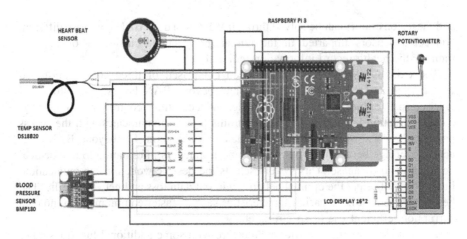

FIGURE 7.8 Configurations of Sensors.

stored in the WBAN controller. The copied values are stored for further reference in a separate data pool. Now, these data must be sent to the medical personnel, and hence, a suitable network is needed based on the requirements. For efficient network selection, the concept of the cognitive radio controller (CRC) has been adopted. The CRC uses the dynamic spectrum access approach to achieve adaptive spectrum allocation and efficient spectrum access. The CRC has been connected using a short-range network with a WBAN controller — some examples are Bluetooth, Wi-Fi, etc. In CRC, the presence of the primary user is monitored. If the primary user is idle, then the CRC allows the cognitive user to use the band until the primary user is in active mode.

6.2 Channel Selection Layer (Tier 2)

This layer is concerned with channel selection for transferring the information to the end-users. There are many networks — such as Wi-Fi, WiMAX, LTE, VoLTE, etc.—available to transmit the data to hospitals. The selection of an optimized channel precedes selecting a suitable channel. in selecting the optimum channel, network attributes such as radio condition, bandwidth, coverage, and handoff latency and user attributes such as residual battery power, speed, power consumption, and QoS parameters have to be considered. In selecting the optimized channel, FIS has been adopted in the proposed system. FIS is the major decision-making logic system that consists of the member function, fuzzy rule set, and surface curve.

The Mamdani fuzzy inference process has been applied through four steps: (1) fuzzification of the input variables, (2) rule evaluation, (3) aggregation of the rule outputs, and, finally, (4) de-fuzzification. Initially, the crisp inputs are taken, and their degrees are determined. These fuzzified inputs are applied to the antecedents of the fuzzy rules. Since the fuzzy rule processes multiple antecedents, the fuzzy operator (AND) has been utilized to get a single number, and it denotes the result of the antecedent evaluation. To evaluate the rule, antecedents AND fuzzy operation intersection have been applied and then, this number (the truth value) is applied to the consequent membership function. The optimization is a simulation where Mamdani and Sugeno methods have been adopted using MATLAB simulation window. After analyzing the channel selection efficiency by using both methods, it is evident that Sugeno provides channel selection with an efficiency of 98%, and Mamdani results in 99% efficiency.

After selecting the identified idle channels, the cognitive users correlate the channel parameters with the end-user network in transferring the information over that network with the DSA approach. The accumulated data are stored in the data pool for further reference, and it acts as a storage house for all medical records. The results can be viewed through an interface where the authority needs to give their personal information. For secure transmission of medical data to the medical personnel, it is transmitted with encryption. The encrypted text is sent to the doctor in a hexadecimal format via e-mail. The proposed algorithm is Base 64, which is a two-way encryption algorithm.

6.2.1 WBAN Controller

The proposed system has been designed with the idea of providing an effective solution to the issues in the medical domain using the latest technology. The process has been initiated from the sensors connected over the body of the patients, and the sensors continuously monitor their physical condition. Hence, three basic sensors—temperature sensors, pulse rate sensors, and blood pressure sensors—have been incorporated. The values of the sensors are rendered in the database in a time span of 10 s, and an e-mail is sent within 30 minutes to the doctor. The data that is collected from these sensors are accumulated in the central node. The sample data in each sensor are shown in Table 7.2.

Hence, each sensor in WSN needs data aggregation functions that include routing the data of all neighbor sensors to a central point, as they may be physically away from the central point; for WBAN, this feature is not required — all the nodes are situated usually at relatively short distances, and they are accessible to the central point. The main intention of this work is to send medical history without any delay or data loss with fast communication. For this purpose, the optimal channel selection in the best network must be chosen to transfer the data. There are many networks available around the premises, but an efficient transmitter is required.

TABLE 7.2
Sample Data Extraction

Patient List	Temperature Sensor	Pulse Rate Sensor	Blood Pressure Sensor
Person A	34.2	65	125
Person B	32.0	72	140
Person C	33.4	63	132
Person D	32.8	78	121
Person E	32.5	86	141

6.2.2 Cognitive Radio Controller

In wireless communication, efficient spectrum usage is the major constraint in using any network to provide efficient results. This part of the implementation work deals with the WBAN controller to extract patient's data through sensors, and the simulation part selects the best network to optimize the selection and secure the data using a two-way encryption algorithm. To overcome the constraints of channel selection for effective DSA, there is an emerging technology called CR which is an adaptive, intelligent radio and network technology. It can detect available channels automatically in a wireless spectrum and can change transmission parameters to enable more communications concurrently. CR also improves the operating behavior of radio communication. The technology of CR involves the dynamic way of accessing the channels in the available spectrum. When the primary user leaves the channel idle, then the SU uses that left-over channel until the primary user comes into focus. The central node is connected with the cognitive controller through short-range wireless networks such as Bluetooth, Wi-Fi, Zigbee, etc. For long-range communication with the data pool, the network with QoS requirements is needed. The CR controller analyses the available network with liable parameters. Since there are many networks—and to overcome the constraint of on-time delivery of information—the selection of network must be optimized. For network selection optimization, FIS has been used to yield efficient results.

6.3 APPLICATION LAYER (TIER 3)

When the patient attains an abnormal condition, the medical authority should immediately be alerted. A database server is necessary to track the previous medical records of the patient for a better diagnosis. The data collected from the sensor is transmitted to a Xamp-based database server to find the record or analyze the patient's data. It will help the doctor make better diagnoses and provide a prescription. The doctor can access the data server at any time, and the doctor can also know the status of the patient's medical condition. The health record of patients is also maintained in the web portal for future use. The portal also possesses the option to maintain and track the records of multiple patients. The patient can also refer to his/her medical details on the web portal. Thus, this system provides an efficient and robust way to maintain and analyze one's medical record.

7 APPLICATION OF DSA IN POST DISASTER MANAGEMENT APPLICATIONS

The Unmanned Aerial Vehicle for Post Disaster Management model illustrated in Figure 7.9 consists of the multicopter with sensors, 64-bit quad-core Cortex – A72 processor, nRF24L01 transceiver in it, and the off-site console. It analyzes different parameters to attain maximum accuracy in dynamic path planning. The cognitive controller with CRN consists of several sensors, a microcontroller, and a communication device which can transfer data at a speed of 2 Mbps using an ISM band of 2.5 GHz—which means it does not need any GSM or GPS systems that may sometimes not function in a disaster-prone zone due to lack of

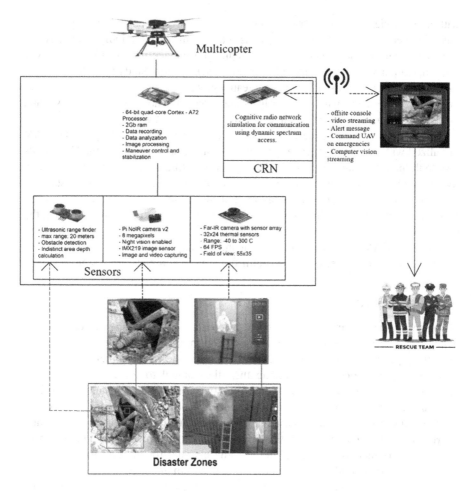

FIGURE 7.9 Post Disaster Management Applications with CRN.

proper tower workings. The microcontroller uses 2GB RAM for quicker processing. Hand in hand analyzing different data in parallel (i.e. processing images of the zone site and on-site processing involves image recognition using computer vision) where two types of images — normal camera image and thermal image — are being processed. The Pi camera image can be used at night through night vision, and a thermal camera is used to capture temperature data of substances or lives.

Both types of images are necessary in different situations as the disaster zone can have many types of environments like submerged, dark, set ablaze, etc. The UAV uses a gyroscope with an accelerometer (MPU6050), which is used to stabilize the UAV when stuck by an external force. The accelerometer is used to maintain the speed in a congested area. An ultrasonic range finder with a range of 100 m is used to detect the range between an obstacle and the UAV. This range data varies from time to time, and a quicker variation depicts that the UAV is going to be in interference with an obstacle, and the path planning method involves escape planning to avoid obstacle interference. This sensor is also used to find the depth in indistinct zones that cannot be seen by the camera. The communication device used can have two-way communication when necessary, like sending backup commands from the console. The off-site console is used for image streaming, which can also be a video. It gives information and alerts when human lives are detected. The status of the UAV and assets are sent as quickly as possible without a loss of image.

The UAV can work in the disaster zone as described below. Deploy the UAV initialized with the area coverage at the disaster affected zone. Then, the UAV finds a suitable path to enter the zone and analyzes it to find obstacles or assets. If any living being is found, the rescue team is informed about its status; if an obstacle is detected, the UAV avoids dashing on it using obstacle avoidance and escape planning. If no living being is detected, then the UAV keeps on analyzing the environment. The UAV checks the coverage area status during its flight. If fully-covered, the UAV returns to the rescue team and lands; otherwise, it continues its flight and analyzes the environment until the area is completely covered.

8 CONCLUSION

In this work, a new approach has been proposed to improve the performance of dynamic spectrum access by identifying the optimum idle channel using the co-operative spectrum sensing approach and selecting the idle channel based on the channel characteristics, and assigning the idle channel to the cognitive user by optimization method. With help of link quality maintenance scheduler algorithm, the transmission flow of CR communication has been maintained. In CRN, the linking and scheduling is done based on parameters such as transmission power and overhead problem reduction. By this approach, the performance of dynamic spectrum access is increased by selecting the idle channel for the cognitive user; the scheduling algorithm also reduces delay in CR communication.

In future works, secure data transmissions, multiple parameter optimization, and collision probability between cognitive users in remote application may be investigated.

REFERENCES

1. Rawat, P., Singh, K. D., and Bonnin, J. M. (2016). Cognitive radio for M2M and Internet of Things: A survey. *Computer Communications*, 94, 1–29.
2. Zhao, H., Ding, K., Sarkar, N. I., Wei, J., and Xiong, J. (2018). A simple distributed channel allocation algorithm for D2D communication pairs. *IEEE Transactions on Vehicular Technology*, 67(11), 10960–10969.
3. Panda, S. K. and Jana, P. K. (2016). Uncertainty-based QoS min–min algorithm for heterogeneous multi-cloud environment. *Arabian Journal for Science and Engineering*, 41(8), 3003–3025.
4. Rajaguru, R. and Vimaladevi, K. (2016). Performance analysis of radio access techniques in self-configured next generation wireless networks. *Advances in Natural and Applied Sciences*, 10(10), 232–242.
5. Prabhananthakumar, M., Rajaguru, R., and Rathnamala, S. (2012). A survey on cognitive engine. *Journal Academic, Industrial & Research*, 1(1).

6. Masonta, M. T., Mzyece, M., and Ntlatlapa, N. (2012). Spectrum decision in cognitive radio networks: A survey. *IEEE Communications Surveys & Tutorials*, *15*(3), 1088–1107.
7. Uyanik, G. S. and Oktug, S. (2017). Cognitive channel selection and scheduling for multi-channel dynamic spectrum access networks considering QoS levels. *Ad Hoc Networks*, *62*, 22–34.
8. Kour, H., Jha, R. K., and Jain, S. (2018). A comprehensive survey on spectrum sharing: Architecture, energy efficiency and security issues. *Journal of Network and Computer Applications*, *103*, 29–57.
9. Thakur, P., Kumar, A., Pandit, S., Singh, G., and Satashia, S. N. (2018). Performance analysis of cognitive radio networks using channel-prediction-probabilities and improved frame structure. *Digital Communications and Networks*, *4*(4), 287–295.
10. Roy, A., Midya, S., Majumder, K., Phadikar, S., and Dasgupta, A. (2017). Optimized secondary user selection for quality of service enhancement of two-tier multi-user cognitive radio network: a game-theoretic approach. *Computer Networks*, *123*, 1–18.
11. Jayakumar, L., and Janakiraman, S. (2019). A novel need based free channel selection scheme for cooperative CRN using EFAHP-TOPSIS. *Journal of King Saud University-Computer and Information Sciences*.
12. Chilakala, S. and Ram, M. S. S. (2018). Optimization of cooperative secondary users in cognitive radio networks. *Engineering Science and Technology, an International Journal*, *21*(5), 815–821.
13. Bhoi, S. K., Panda, S. K., Patra, B., Pradhan, B., Priyadarshinee, P., Tripathy, S., ... Khilar, P. M. (2018, December). FallDS-IoT: A fall detection system for elderly healthcare based on IoT data analytics. In *2018 International Conference on Information Technology (ICIT)* (pp. 155–160). IEEE.
14. Bhoi, S. K., Panda, S. K., Padhi, B. N., Swain, M. K., Hembram, B., Mishra, D., ... Khilar, P. M. (2018, December). Fireds-iot: A fire detection system for smart home based on iot data analytics. In *2018 International Conference on Information Technology (ICIT)* (pp. 161–165). IEEE.
15. Bayrakdar, M. E. (2020). Cooperative communication based access technique for sensor networks. *International Journal of Electronics*, *107*(2), 212–225.
16. Oyewobi, S. S., Hancke, G. P., Abu-Mahfouz, A. M., and Onumanyi, A. J. (2019). A delay-aware spectrum handoff scheme for prioritized time-critical industrial applications with channel selection strategy. *Computer Communications*, *144*, 112–123.
17. Cicioğlu, M., Bayrakdar, M. E., and Çalhan, A. (2019). Performance analysis of a mew MAC protocol for wireless cognitive radio networks. *Wireless Personal Communications*, *108*(1), 67–86.
18. Bayrakdar, M. E. and Çalhan, A. (2018). Artificial bee colony–based spectrum handoff algorithm in wireless cognitive radio networks. *International Journal of Communication Systems*, *31*(5), e3495.
19. Wang, H., Yao, Y. D., and Peng, S. (2018). Prioritized secondary user access control in cognitive radio networks. *IEEE Access*, *6*, 11007–11016.

8 TB-PAD

A Novel Trust-Based Platooning Attack Detection in Cognitive Software-Defined Vehicular Network (CSDVN)

Rajendra Prasad Nayak[1], Srinivas Sethi[2], and Sourav Kumar Bhoi[3]

[1]Department of Computer Science and Engineering, Government College of Engineering, Kalahandi, India, Email: rajendra.cet07@gmail.com
[2]Deparment of Computer Science Engineering and Applications, Indira Gandhi Institute of Technology, Sarang, India, Email: srinivas_sethi@igitsarang.ac.in
[3]Department of Computer Science and Engineering, Parala Maharaja Engineering College, Berhampur, India, Email: sourav.cse@pmec.ac.in

1 INTRODUCTION

Vehicular technology is nowadays booming for providing safety and non-safety applications to the users [1–4]. As accidents are growing every day, it is a problem for the government and private organizations to save the life of humans. Therefore, vehicular ad hoc network (VANET) provides a platform of manual, semi-automated, and fully-automated technology. By using the technologies, the vehicles in the network communicate essential information with neighbors, other vehicles, RSUs, BS, local controller, and SDN controller. For this communication, vehicles use vehicle to vehicle (V2V) and vehicle to infrastructure communication (V2I). Vehicles in the network are highly dynamic; therefore, the vehicles use position-based routing to transfer the information to other nodes for reliable communication.

Nowadays, Software Defined Network (SDN) provides a better architecture for VANET, which improves the performance of the network [5–8]. SDN is a centralized architecture that divides the network into a data plane and control plane for services like QoS, security, optimization, low cost, availability, and better resource utilization in a dynamic environment. It makes the network flexible, where the total control of the network is with the SDN controller at the control end. From the control end, the SDN controller controls the whole network by setting rules and regulations. The two layers are connected by the Openflow protocol. The two technologies are hybridized to generate a Software-Defined Vehicular Network (SDVN). The SDVN network divides the network into the data plane and control plane. In the control plane, the SDN controller exists and, in the data plane, the vehicle exists but has small intelligence for processing data, and it follows the policies set by SDN to execute all works. This will be a better technology to divide the work into both parts. However, if the vehicles use cognitive computing technology, then it will be better for the network to perform well. Cognitive computing [9–11] facilitates the SDVN to work better by using all techniques such as AI, machine learning, reasoning, etc. The techniques help predict, classify, etc. by taking the prior knowledge about the network behavior. This concept motivates us to build a CSDVN network.

Security is a major concern in urban VANETs as many types of attacks are possible nowadays—the attackers try to control the network and reduce network performance by disturbing the network applications [12–42]. A platooning attack [12,13] is such a type of attack where a group of vehicles in the network work together to affect the network activities. They move in a group where a group leader (leader vehicle) is present with other vehicles (follower vehicles). This attack needs to be detected in the network.

The major contributions in this work are stated as follows:

1. A novel TB-PAD method is proposed for CSDVN. In this method, the RSUs at the junctions receive the beacon information for the detection of the platoon. The information is sent to the LC through the BS for the identification of suspected vehicles in the platoon.
2. The platoon information is sent to the RSUs as well as the SDN controller by the LC for the update of trust values. The values are broadcasted by the RSUs to the nearby vehicles for update of trust values at their neighbor table.
3. This method mainly uses cognitive computing technology which uses prior knowledge of the vehicles to generate the new trust values.
4. The method is evaluated using OMNET++ by considering various network parameters.

The rest of the work is distributed as follows. Section 2 presents the TB-PAD method. Section 3 presents related works. Section 4 presents the simulation and results. Section 5 presents the conclusion and future scope.

2 TB-PAD PROPOSED METHODOLOGY

This section describes the network model, the misbehavior model, and the proposed approach. The network model describes network devices and the communication between them, shown in Figure 8.1. The misbehavior model describes the

| source Vid | position | destination | direction | speed | trust value |

FIGURE 8.1 Packet format for beacon.

misbehavior performed by attacker vehicles in the network. The proposed method discusses the platooning attack detection method in CSDVN.

3 RELATED WORKS

Many works have been done to detect the platooning attack in the network. However, in this work, we have used the CSDVN network which gives freshness to the whole work. The attack in this network is also checked using our TB-PAD mechanism. Researchers mainly work in SDVN, VANET, MANET, WSN, IoT, etc. to detect the group attacks. The platooning attack is very specific here and very little work has been done by the researchers to detect this attack in SDVN. Many such related works are presented in [12–42].

3.1 Network Model

1. **Vehicle:** The network consists of vehicles (V) with one board unit (OBU) in the city area. The vehicle is able to send, receive, and route packets. It has processing, storage, and intelligence capability. It has high battery status. The vehicle can communicate with other vehicles or RSU using V2V and V2I communication, using wireless links in an ad hoc fashion. A vehicle beacons its current position, destination, direction, speed, and trust value in a particular period in its communication range. The current position is calculated using GPS and maps at the vehicle OBU unit. The beacon structure is shown in Figure 8.2. The vehicle has a neighbor table where it updates the data it receives from another vehicle. A vehicle can be attacked by another vehicle or a group of vehicles. Vehicles use a position-based routing protocol to send the data to a destination.
2. **RSU:** RSU can send, receive, and route the packets to the destination. RSU is set at the junction to record the information (beacon) of the vehicles moving from one junction to another. It has processing, storage, and intelligence capability. It has high power status. RSU communicates with the vehicles using I2V communication and it communicates with the BS using Wimax/LTE connection. RSU stores the information of all vehicles in the RSU table. The RSU table consists of the same updated data as in the neighbor table of the vehicle. RSU is a genuine node. It forwards the RSU table to the BS in a particular interval of time. The RSU table also consists of specific types of message a vehicle send (safety or non-safety).
3. **BS:** The BS is responsible for receiving and transferring data from one layer to another (data plane to control plane and vice versa). It has processing, storage, and intelligence capability. It has high power status.

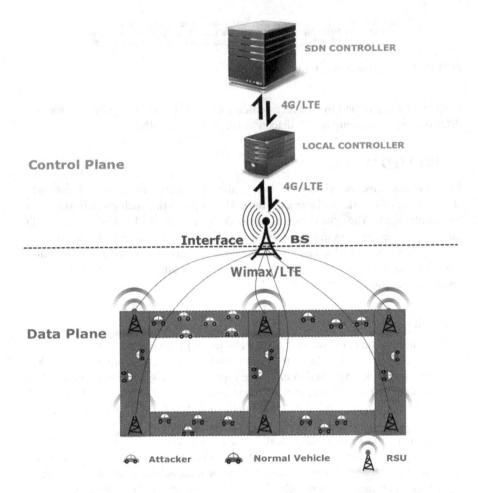

FIGURE 8.2 Proposed TB-PAD Method Framework to Detect the Platooning Attack in CSDVN.

4. **Interface:** Interface is the layer that connects the two planes by a BS. The BS acts as an interface that transfers data packets and control packets between layers. It also sends flow rules and policy rules from the control layer to the data layer.
5. **LC:** The LC at the control plane is responsible for processing, receiving, and transferring data to the SDN controller or BS. It has processing, storage, and intelligence capability. It has high power status. LC helps SDN by performing smaller tasks and sending bigger tasks to the SDN controller. Here, we consider the trust computation as a smaller task performed at LC. LC also stores the previous trust values which are updated.
6. **SDN Controller:** The SDN controller at the control plane runs software that manages all the work in the CSDVN network. It defines the above policies and rules, and flow rules for the entire network devices in the network. It has

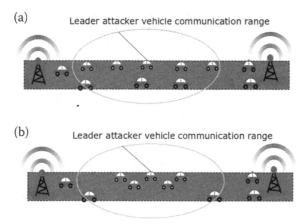

FIGURE 8.3 Platooning Attack by the Attacker Vehicles in CSDVN.

high processing, high storage, and high intelligence capability. It has high power status. SDN stores the platoon attack group information which is updated.

3.2 Misbehavior Model

In this work, we have mainly focused on a misbehavior performed by the vehicles in the CSDVN network. The vehicles perform a platooning attack. In this, the vehicles in the network move together and disturb the network in any sense. They may do jamming, packet dropping, distributed denial of service attack (DDOS), ID spoofing, position cheating, etc. The group consists of a leader vehicle and follower vehicles; the follower vehicles follow the leader vehicle. Figure 8.3 (a) and (b) show the vehicles in the group attacking an arbitrary pattern. The vehicles are moving in a line; the leader attacker vehicle may be in middle, last, or first, but all followers should be the neighbors. The vehicles may also arrange in an arbitrary fashion such as the leader attacker vehicle will be at the centroid, or any corner but all follower vehicles are neighbors. They move in the same direction, same destination, at almost the same speed, and send the same type of messages (alert messages, safety, and non-safety messages)—all activities are the same. The rules are defined by the leader's vehicle.

3.3 Proposed Methodology for TB-PAD

In this section, we described the method and algorithm of the whole network. All vehicles beacon a short message to its neighbors and a neighbor table is maintained. When the vehicles move through a junction, the beacons are captured by the RSU

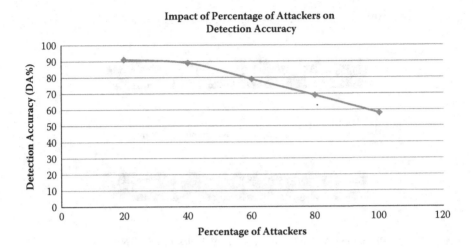

FIGURE 8.4 Detection Accuracy.

stored in the RSU table. Then, it sends the table data to the BS. The BS again forwards the table data to the LC. The LC receives the data from all RSUs. That means it receives all city vehicle information at a particular time. After receiving all vehicle data, it finds the vehicles that are common neighbors to each other. They also have the same direction, same destination, almost at the same speed (same range), and sends any safety and non-safety messages. The safety and non-safety messages are also specific type. LC finds the neighbor table from the RSU table and checks the similarity between vehicles by considering all the above parameters, the LC tags a trust value (1/0; 0=suspected,1=not suspected) to a vehicle and generates the suspected vehicles. The similarity matrix table consists of the trust values. The table is shown in Table 8.1; here, V3, V4, and V5 are the vehicles of neighbors of each other and their actions are the same with respect to the above parameters. The trust value is calculated as follows:

$$Tcurrent = \alpha \times Tcurrent + (1 - \alpha) \times Tprevious \qquad (8.1)$$

where α and $(1 - \alpha)$ are 0.5 each, Tcurrent is the current trust value and Tprevious is the previous trust value at the LC. After creating similarity tables for each RSU, the LC matches the same patterns of vehicles in at least th tables. If those patterns are present, then these groups of people with a group ID are detected as suspected group members (SGM), and this information is forwarded to SDN as well as RSU. The RSU broadcasts the SGM information in the network to update the trust values at the vehicle level. The algorithm is shown in Algorithm 1 as follows.

> **Algorithm 1 Pseudocode for Detection of Suspected Group in CSDVN.**
>
> **Input:** RSU table (vid, direction, speed, destination, speed, specific message type)
> **Output:** Suspected Group Member
>
> 1. 1 **For** RSU = 1 to r
> 2. **For** n = 1 to p
> 3. **For** m = 1 to p
> 4. 4 **If** (V[m] is neighbor && V[m] direction is same && V[m] destination is same && V[m] speed range is same && V[m] message type is same)
> 5. Set Tcurrent for V[m] = 0; // suspected
> 6. Calculate Tcurrent;
> 7. Store in Similarity table of RSU[r];
> 8. **Else**
> 9. Set Tcurrent for V[m] = 1; //not suspected
> 10. Calculate Tcurrent;
> 11. Store in Similarity table of RSU[r];
> 12. **IfEnd**
> 13. **End For**
> 14. **EndFor**
> 15. **EndFor**
> 16. **If** (same pattern of vehicles are present in th similarity tables)
> 17. Set Suspected Group id;
> 18. V = Suspected Group Member (SGM) for that id;
> 19. Send group id and SGM information to RSU and SDN controller;
> 20. RSU broadcast SGM information;
> 21. Vehicle updates trust values of neighbors;
> 22. **EndIf**

4 SIMULATION AND RESULTS

In this section, the simulation and results are presented. Initially, the simulation setup and assumptions taken during the simulation are discussed. The parameters considered for performance evaluation are detection accuracy, detection time, and energy consumption.

To carry out the simulation, we have used the Veins hybrid framework simulator. This simulator uses the IEEE 802.11p standard for communication. Since the framework is a hybrid one, it uses OMNeT++ and Simulation of Urban Mobility

TABLE 8.1
Similarity Matrix Trust (0/1) Table of RSU[1]

	V1	V2	V3	V4	V5
V1	0	0	0	1	0
V2	1	0	0	0	1
V3	0	0	0	1	1
V4	1	0	1	0	1
V5	0	0	1	1	0

[1] Vehicles V3, V4, and V5 Show Similar Actions and Are Suspected of Performing a Platooning Attack.

(SUMO) as network and road traffic simulator, respectively [43]. These simulators are integrated using a Traffic Control Interface (TraCI). This interface provides the TCP connection between the network and road traffic simulator and maintains real-time interaction between them. We simulate our work in a Grid Map scenario. The configuration parameters for SUMO and OMNeT++ are provided in Tables 8.2 and 8.3, respectively. The performance is evaluated by varying the malicious activity in the network. The percentage of malicious nodes is varied from 20-100% (16, 32, 48, 64, 80) of 80% malicious nodes of 100 vehicles. The group size for a platooning attack is four nodes. For example, in the case of 16 nodes, four groups are made of size 4. The group members are common neighbors of each other, and the group is randomly deployed. The other 84 nodes are also randomly deployed in the network. The initial trust value of malicious vehicles is set to 0.5 and for genuine node, it is 0.8. The results are the average of 20 simulation runs.

From Figure 8.4, it is observed that the detection accuracy reduces with the increase of malicious vehicles in the network. However, it shows 91%, 89%, 79%, 69%, and 58% when the malicious vehicles are 20%, 40%, 60%, 80%, and 100%, respectively. From Figure 8.5, it is observed that the energy consumption increases with the increase of malicious vehicles in the network. However, it shows 18.3, 20.7, 23, 26.1, and 30 J when the malicious vehicles are 20%, 40%, 60%, 80%, and 100%, respectively. From Figure 8.6, it is observed that the detection time increases with the increase of malicious vehicles in the network. However, it shows 35, 43, 53, 67, and 89 seconds when the malicious vehicles are 20%, 40%, 60%, 80%, and 100%, respectively. From the above discussion, it is seen that the proposed method performs well in detecting the platooning attack in all parameter level conditions.

5 CONCLUSION

From the above work, it is found that the proposed TB-PAD method can detect the attacker vehicles with better accuracy. The detection time is also less in detecting

TABLE 8.2
Simulation Setting for SUMO

Sl. No.	Parameters	Values
1	Area	3000 × 2000 m^2
2	Number of lanes	3 × direction
3	Maximum speed of vehicles	30 m/s
4	Allowed maximum speed on edges	30.556 m/s
5	Maximum acceleration	3.0 m/s^2
6	Maximum deceleration	6.0 m/s^2
7	Driver imperfection	0.5
8	Number of vehicles	100
9	Vehicles type	3
10	Length of vehicle type	5 m, 7 m, 12 m
11	Speed of malicious vehicles in group	20 m/s
12	Destination of malicious vehicles in group	Same
13	Message sends by malicious vehicles	Specific alert message taken

TABLE 8.3
Simulation Setting for Network Simulator

Sl. No.	Parameters	Values
1	Simulation time	300 s
2	Bitrate	6 Mbps
3	Packet generation rate	10 packets/s
4	Communication range of vehicle	300 m
5	Communication range of RSU	500 m
6	Update interval	0.1s
7	IEEE	802.11p
8	Sensitivity	−80 dBm
9	Th	3
10	SDN controller	1
11	Local controller	1
12	Number of RSU	4
13	BS	1
14	Number of simulations	20

malicious nodes. The energy consumption of the nodes is found to be more for malicious vehicles as they are involved in different activities. The TB-PAD will be a better algorithm to detect platooning attacks in WSN, VANET, SDVN, IoT, etc.

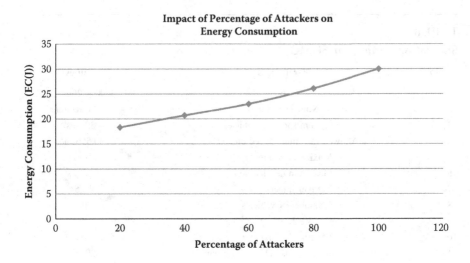

FIGURE 8.5 Energy Consumption.

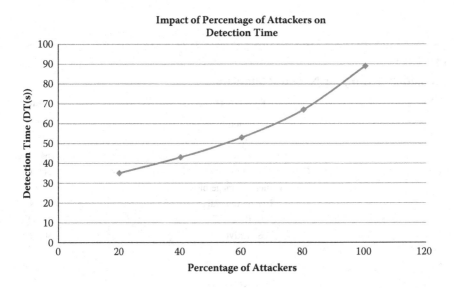

FIGURE 8.6 Detection Time.

networks. Simulation works show that this method can be implemented practically in a city area. In the future, we will compare this method with some standard platooning attack detection schemes. We will implement this work using application-based software and hardware for better validation.

REFERENCES

1. Hasrouny, H., Samhat, A. E., Bassil, C. and Laouiti, A. (2017). VANet security challenges and solutions: A survey. *Vehicular Communications 7*, pp. 7–20.
2. Li, F. and Wang, Y. (2007). Routing in vehicular ad hoc networks: A survey. *IEEE Vehicular Technology Magazine 2*(2), pp. 12–22.
3. Cunha F., Boukerche, A., Villas, L., Viana, A., and Loureiro, A. A. (2014). Data communication in VANETs: a survey, challenges and applications. *Network*, IEEE Communications Surveys and Tutorials.
4. Bhoi, S. K. and Khilar, P. M. (2013). Vehicular communication: A survey. *IET Networks 3*(3), pp. 204–217.
5. Kreutz, D., Fernando M. V. R., Verissimo, P. E., Rothenberg, C. E., Azodolmolky, S., and Uhlig, S. (2014). Software-defined networking: A comprehensive survey. *Proceedings of the IEEE 103*(1), pp. 14–76.
6. Ku, I., Lu, Y., Gerla, M., Gomes, R. L., Ongaro, F., and Cerqueira, E. (2014). Towards software-defined VANET: Architecture and services. In *2014 13th Annual Mediterranean Ad Hoc Networking Workshop (MED-HOC-NET)* (pp. 103–110). IEEE.
7. Truong, N. B., Lee, G. M., and Ghamri-Doudane, Y. (2015). Software defined networking-based vehicular adhoc network with fog computing. In *2015 IFIP/IEEE International Symposium on Integrated Network Management (IM)* (pp. 1202–1207). IEEE.
8. Arif, M., Wang, G., Wang, T., and Peng, T. (2018). SDN-based secure VANETs communication with fog computing. In *International Conference on Security, Privacy and Anonymity in Computation, Communication and Storage* (pp. 46–59). Springer, Cham.
9. Marfia, G., Roccetti, M., Amoroso, A., Gerla, M., Pau, G., and Lim, J.-H. (2011). Cognitive cars: Constructing a cognitive playground for VANET research testbeds. In Proceedings of the 4th International Conference on Cognitive Radio and Advanced Spectrum Management (pp. 1–5).
10. Abuelela, M., and Olariu, S. (2010). Taking VANET to the clouds. In *Proceedings of the 8th International Conference on Advances in Mobile Computing and Multimedia* (pp. 6–13).
11. Modha, D. S., Ananthanarayanan, R., Esser, S. K., Ndirango, A., Sherbondy, A. J., and Singh, R. (2011). Cognitive computing. *Communications of the ACM 54*(8), pp. 62–71.
12. Sakiz, F. and Sen, S. (2017). A survey of attacks and detection mechanisms on intelligent transportation systems: VANETs and IoV. *Ad Hoc Networks 61*, pp. 33–50.
13. Kerrache, C. A., Calafate, C. T., Cano, J.-C., Lagraa, N., and Manzoni, P. (2016). Trust management for vehicular networks: An adversary-oriented overview. *IEEE Access 4*, pp. 9293–9307.
14. Hu, H., Lu, R., Zhang, Z., and Shao, J. (2016). REPLACE: A reliable trust-based platoon service recommendation scheme in VANET. *IEEE Transactions on Vehicular Technology 66*(2), pp. 1786–1797.
15. Lyamin, N., Vinel, A., Jonsson, M., and Loo, J. (2013). Real-time detection of denial-of-service attacks in IEEE 802.11 p vehicular networks. *IEEE Communications Letters 18*(1), pp. 110–113.
16. Hu, H., Lu, R., Huang, C., and Zhang, Z. (2016). Tripsense: A trust-based vehicular platoon crowdsensing scheme with privacy preservation in vanets. *Sensors 16*(6), p. 803.
17. Amoozadeh, M., Deng, H., Chuah, C.-N., Zhang, H. M., and Ghosal, D. (2015). Platoon management with cooperative adaptive cruise control enabled by VANET. *Vehicular Communications 2*(2), pp. 110–123.

18. Boeira, F., Marinho, P. B., de Freitas, E. P., Vinel, A., and Asplund, M. (2017). Effects of colluding Sybil nodes in message falsification attacks for vehicular platooning. In 2017 IEEE Vehicular Networking Conference (VNC) (pp. 53–60). IEEE.
19. Asplund, M. (2014). Poster: Securing vehicular platoon membership. In 2014 IEEE Vehicular Networking Conference (VNC) (pp. 119–120). IEEE.
20. Patounas, G., Zhang, Y., and Gjessing, S. (2015). Evaluating defence schemes against jamming in vehicle platoon networks. In 2015 IEEE 18th International Conference on Intelligent Transportation Systems (pp. 2153–2158). IEEE.
21. Vitelli, D. (2016). Security vulnerabilities of vehicular platoon network. *University of Naples Federico II Studies, Thesis*.
22. Mutaz, A., Muhammad, L. M., and Chellappan, S. (2013). Leveraging platoon dispersion for sybil detection in vehicular networks. In 2013 Eleventh Annual Conference on Privacy, Security and Trust (pp. 340–347). IEEE.
23. Lyamin, N., Kleyko, D., Delooz, Q., and Vinel, A. (2018). AI-based malicious network traffic detection in VANETs. *IEEE Network 32*(6), pp. 15–21.
24. Zhang, T. and Zhu, Q. (2018). Distributed privacy-preserving collaborative intrusion detection systems for VANETs. *IEEE Transactions on Signal and Information Processing over Networks 4*(1), pp. 148–161.
25. Kumar, N. and Chilamkurti, N. (2014). Collaborative trust aware intelligent intrusion detection in VANETs. *Computers & Electrical Engineering 40*(6), pp. 1981–1996.
26. Lal, A. S. and Nair, R. (2015). Region authority based collaborative scheme to detect Sybil attacks in VANET. In 2015 International Conference on Control Communication & Computing India (ICCC) (pp. 664–668). IEEE.
27. Zhang, J. (2011). A survey on trust management for vanets. In 2011 IEEE International Conference on Advanced Information Networking and Applications (pp. 105–112). IEEE.
28. Mejri, M. N., Ben-Othman, J., and Hamdi, M. (2014). Survey on VANET security challenges and possible cryptographic solutions. *Vehicular Communications* 1(2), pp. 53–66.
29. Engoulou, R. G. (2014). Martine Bellaïche, Samuel Pierre, and Alejandro Quintero. "VANET security surveys." *Computer Communications 44*, pp. 1–13.
30. Yan, G., Olariu, S., and Weigle, M. C. (2008). Providing VANET security through active position detection. *Computer Communications* 31(12), pp. 2883–2897.
31. Arif, M., Wang, G., Wang, T., and Peng, T. (2018). SDN-based secure VANETs communication with fog computing. In *International Conference on Security, Privacy and Anonymity in Computation, Communication and Storage* (pp. 46–59). Springer, Cham.
32. Todorova, M. S. and Todorova, S. T. (2016). DDoS Attack Detection in SDN-based VANET Architectures. June, 175.
33. Bhoi, S. K. and Khilar, P. M. (2013). A secure routing protocol for vehicular ad hoc network to provide ITS services. In 2013 International Conference on Communication and Signal Processing (pp. 1170–1174). IEEE.
34. Bhoi, S. K. and Khilar, P. M. (2014). SIR: A secure and intelligent routing protocol for vehicular ad hoc network. *IET Networks* 4(3), pp. 185–194.
35. Bhoi, S. K. and Khilar, P. M. (2016). RVCloud: A routing protocol for vehicular ad hoc network in city environment using cloud computing. *Wireless Networks* 22(4), pp. 1329–1341.
36. Swain, R. R., Khilar, P. M., and Bhoi, S. K. (2018). Heterogeneous fault diagnosis for wireless sensor networks. *Ad Hoc Networks* 69, pp. 15–37.
37. Bhoi, S. K. and Khilar, P. M. (2012). SST: A secure fault-tolerant smart transportation system for vehicular ad hoc network. In *2012 2nd IEEE International Conference on Parallel, Distributed and Grid Computing* (pp. 545–550). IEEE.

38. Bhoi, S. K. (2013). SGIRP: A Secure and Greedy Intersection-Based Routing Protocol for VANET using Guarding Nodes. *PhD diss.*
39. Bhoi, S. K. and Khilar, P. M. (2016). Self soft fault detection based routing protocol for vehicular ad hoc network in city environment. *Wireless Networks 22*(1), pp. 285–305.
40. Bhoi, S. K., Nayak, R. P., Dash, D., and Rout, J. P.(2013). RRP: A robust routing protocol for vehicular ad hoc network against hole generation attack. In 2013 International Conference on Communication and Signal Processing (pp. 1175–1179). IEEE.
41. Nayak, R. P., Sethi, S., and Bhoi, S. K. (2018). PHVA: A position based high speed vehicle detection algorithm for detecting high speed vehicles using vehicular cloud. In 2018 International Conference on Information Technology (ICIT) (pp. 227–232). IEEE.
42. Swain, R. R., Khilar, P. M., and Bhoi, S. K. (2018). Heterogeneous fault diagnosis for wireless sensor networks. *Ad Hoc Networks 69*, pp. 15–37.
43. Varga, A. and Hornig, R. (2008). An overview of the OMNeT++ simulation environment. In Proceedings of the 1st International Conference on Simulation Tools and Techniques for Communications, Networks and Systems and Workshops (p. 60). ICST (Institute for Computer Sciences, Social-Informatics and Telecommunications Engineering).

9 Analysis of Security Issues in IoT System

Likhet Kashori Sahu, Sudibyajyoti Jena, Sambit Kumar Mishra, and Sonali Mishra
Department of Computer Science and Engineering, Siksha 'O' Anusandhan Deemed to be University, Bhubaneswar, Odisha, India

1 INTRODUCTION

IoT is a fast-growing technology that consists of physical devices connected to the internet to make existing technologies smarter. It arose from our need to receive more data, control stuff, automate, and finish tasks faster. The physical devices are the "things" that collect, analyze, and process data. Thus, IoT is a technology that provides us with remote access to other devices that can perform a task devoid of human-to-human or human-to-computer interaction. The main vision of IoT is to connect all devices with the internet to form a smart technology that will increase the functionalities of all existing technologies and provide better solutions to current problems.

Over time, IoT has found uses in many spheres of our life like agriculture, industries, transportation, smart houses, and smart cities. IoT has significantly improved the quality of life. Because of its vast applications in various sectors, it's is necessary to secure its framework. The hardware, software, and connecting network should be safeguarded against all security threats and make it reliable and secure for the clients. IoT can expand and become part of our daily lives only if it is made secure and consumer-friendly. IoT, while being a revolutionary technology, is still held back by limited power and energy constraints. To mitigate this problem, the field of Green Internet of Things was proposed. GIoT merges green computing techniques with traditional IoT technologies to make IoT more energy-efficient and optimize its functionality using a minimum amount of resources.

This chapter will begin with a brief review of the IoT technology and the IoT ecosystem. In Section II, we will look into the different layers of IoT architecture (i.e. perception layer, network layer, processing layer, and application layer). Section III will give an understanding of some technologies used in IoT. In Section III, we will discuss the application of IoT in different sectors. Section V will elaborate on security challenges like communication security, cryptography, data protection, and various challenges based on previous researches done on IoT Security. Section VI will provide a list of all possible attacks over the four layers, as mentioned before. Finally, we will conclude this chapter in Section VII.

2 IOT ARCHITECTURE

This section gives details about how IoT architecture works and how data flow takes place from layer to layer. Researchers proposed various model (three-layer model, four-layer model, and so on) depending on their requirement and the task needed to be done. Here, we give a detailed four-layer approach of IoT architecture—perception layer, application layer, network layer, processing layer. This approach will study how raw information is collected and then changed to processed data, analyzed and stored in the database and is used by authorized user [1].

2.1 PERCEPTION LAYER

This is also known as the object layer or recognition layer and is the lowest layer in IoT architecture which represents the physical nature of the IoT technology. It acts as a connection between the physical and virtual world. The main function of his layer is to sense, collect, and modify data from the environment and transfer it to the network layer by selecting a suitable frequency.

2.2 NETWORK LAYER

This is also called the transport layer and has the function of transmitting information from perception layer to other layers [2]. It consists of two sublayers, an encapsulation layer that converts data into small packets, and routing layer that transports packets from source to destination.

2.3 PROCESSING LAYER

This is also known as the support layer. The functionality of the support layer includes mass processing of data and storing it in a data base. This layer act as a communication path between the network layer and application layer. Mostly, the security of this layer depends on the node to node communication path as it does not support security protocols like user authentication and key management [3–5].

2.4 APPLICATION LAYER

This layer provides service based on the end-user's (can be machine or human) need. The application layer provides a service platform for different sectors like healthcare, communication, industrial automation, etc. At present, the two most used IoT application layer protocols are CoAP (Constrained Application Protocols) and CoAP (Pub/Sub) (Constrained Application Protocol Publish\Subscribe). CoAP with low overhead is for communication among resource-constrained devices (like Wireless Sensor Node (WSN)) [4,6].

This IoT protocol plays two logical roles as client and server, where the client send a resource retrieval request to the server that acts as a resource host. Similarly, CoAP Pub/Sub has three roles—client, server, and broker [6] (Figures 9.1–9.4).

Analysis of Security Issues in IoT System

FIGURE 9.1 Dataflow in IoT.

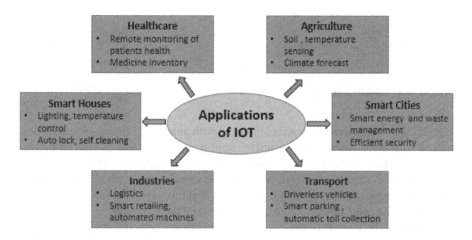

FIGURE 9.2 Some Applications of IoT.

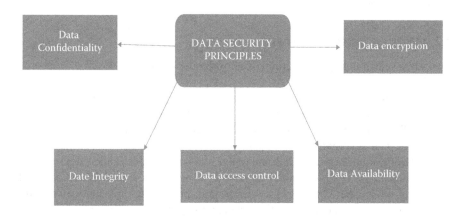

FIGURE 9.3 Principles of Data Security

3 SOME IMPORTANT TECHNOLOGIES

3.1 Radio Frequency Identification

RFID [7] is a technology used by the perception layer to identify things. A RFID system consists of two components [8]: a RFID tag and a RFID reader.

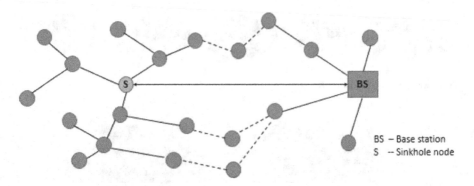

FIGURE 9.4 Sinkhole Attack.

RFID tags are generally a microchip paired with antennae. Objects of interest are marked with three types of tags: passive, semi-passive, and active. Tags can store information and transfer it to the reader in the form of feedback signals.

RFID reader consists of a signal generator and signal receiver whose functions are coordinated by a microcontroller. RFID reader continuously transmits radio waves that are received by objects containing RFID tags. It also receives the feedback transmitted by the tags. This transmitting and receiving of radio waves form a virtual perception that helps in identification and collection of data from the physical world.

On the basis of application [9], there are the different frequencies ranges in RFID: low frequency (<=135 kHz), high frequency (13.56 MHz), and ultra-high frequency (862–928 MHz)

3.2 Wireless Sensor Network

A wireless sensor network is a group of sensor nodes densely deployed over an area called
sensor field, which works together to monitor the environment and collect data such as temperature, pressure, vibration, acceleration etc. The important components of WSN modules [10] include hardware, communication stack, middleware, and secure data aggregation.

A network of nodes collect data and report it to special nodes called sinks using a multi-hop path. It is bidirectional in nature. A sensor is a low-powered, small computing device that consists of a sensing unit, processing unit, transceiver unit, and a power unit. It generally senses in analogue signals that are converted to digital signals before transferred to sink. WSN has a higher range as compared to RFID.

3.3 Green IoT

Nowadays, researchers not only focus on security issues of IoT devices but also on how to minimize the effect they have on the environment. Suppose a sensor node uses a small amount of energy, when thousands of sensor node work together, they will use a huge amount of electrical energy. In the field of green IoT, green computing (GC) and green devices work together to reduce the impact of devices on the environment. The green devices are energy-efficient devices while green computing deals with improving the processing power of these devices while keeping their energy consumption low. GC also focuses on enhancing the security algorithm so they consume less energy.

4 APPLICATIONS OF IOT

In the last decade, IoT has expanded into many fields. The IoT has a wide range of applications in different fields like healthcare, smart cities, agriculture, industrial automation and transport. Integration of IoT into these fields will give us smarter solutions to existing problems.

4.1 IoT in Healthcare

Recently, IoT has redefined healthcare. Wearable sensors in patients constantly track their body function like heart rate, blood pressure, temperature etc. and allow doctors to remotely monitor patient's health. S. M. Riazul Islam et al. [11] in their paper have described an Internet of Things healthcare network (IoThNet) that will lead to smarter hospitals. Prosanta Gope and Tzonelih Hwang in their paper [12] described the technology of Body Sensor Networks (BSNs) that use tiny light-weight sensor nodes to continuously monitor the health parameters of a patient.

IoT has revolutionized the service aspect of healthcare. In [13], the authors have proposed a fire detection system that uses IoT technology with different sensors including carbon dioxide sensor, carbon monoxide sensor, and temperature sensor. In [14], the authors have merged IoT technology and Arduino uno with different low power sensors like accelerometer and gyroscope to form a wearable fall detection system for the elderly patients. These detection systems help save precious lives by promptly notifying the responsible organizations like fire department and hospitals.

Now, the patient-doctor experience is not limited to only scheduled meeting and appointments. Due to these smarter healthcare solutions, more lives are being saved and the cost of healthcare is also reduced.

4.2 IoT in Transport

Implementation of IoT technology in the field of transportation has shown tremendous results. IoT has enabled smart traffic control, smart parking, automatic toll collection, and road assistance. IoT can be used to keep track of precious cargo being transferred and also monitor the driver's health. Sensors on vehicles can also gather information about road and climatic conditions. These smart vehicles can also communicate among themselves to facilitate safe travel. Thus, IoT can solve problems like traffic and road accidents.

4.3 IoT in Smart Houses

Incorporating IoT technology with our home environment will make it more safe, secure, and automated. Motion sensors can detect when someone enters the house and notify us. Internal sensors can control lighting and temperature according to outer climatic conditions. IoT can connect home appliances with our mobile so that we can control them remotely. A smart refrigerator can manage inventory of our food supplies. A smart bed can provide health assistance to elderly members.

4.4 IoT in Agriculture

Due to the recent exponential growth in population, there has been an increase in demand of food supplies. Traditional agricultural methods can't keep up with the rising demand. Thus, modern agriculture techniques need to more efficient. IoT has helped many countries increase their productivity. Sensors are used to monitor the quality of soil, temperature, and climatic conditions. Automated drones can be used to spray fertilizers and pesticides. Using IoT, we can also minimize wastage of resources and reduce human error.

4.5 IoT in Industries

Modern industries have embraced the technology of IoT to increase their productivity. Smart automated machines can work more efficiently than human labor. IoT can also be used to manage inventory of all products.

4.6 IoT in Education

Inclusion of IoT in education scenario can enhance both the experience of teachers and students. Digital learning can complement the physical learning at campus. IoT can also be used to make an attendance tracking system to save time and reduce human error. The university campus can also benefit from IoT by utilizing IoT-enabled efficient parking and security systems [15].

4.7 IoT in Smart Cities

Recently, there has been a shift of population from rural areas to urban cities. Thus, to manage the increase in population, the cities should be efficient and smart. The integration of all applications given above with a city's ecosystem provides solutions to some of the problems faced by modern cities. External sensors can be used to detect air and water pollution. Automated robots and drones can be used to deliver goods. Street surveillance cameras connected with IoT network will ensure the safety of the residents; fast action against any disasters can be taken. Green computing techniques will provide efficient resource allocation and energy management. With the whole of the population connected with a singular IoT network, they can have a say in major decisions regarding the city.

5 SECURITY AND AUTHENTICITY OF DATA IN IOT

The growth in IoT, though it offers a lot of opportunities, still suffers from various security threats. As every day, huge devices of various types are connecting, this makes IoT heterogeneous in nature. As new devices added have diverse security measure, it leads to new security challenges [16].

In this section, we will discuss various security and privacy challenges in the field of IoT. According to [17], security can be defined as protecting IoT devices from any physical damage and unauthorized access within or outside the device. As IoT connects, devices of various types that have different memory capacity, processing power, and energy consumption must have security protocols that are compatible.

IoT devices are not only accessed by the users but also by other devices. In [17], M. Abomhara emphasized that an IoT device must identify that the person or device trying to get access to the service is authorized, and only then is the entity given access to that service. In this field, trust plays an important role; trust is important when different entities connect and share data in an IoT environment. In [18] Brauchli explains that a device is said to be trustworthy if it acts as expected even in a hostile environment. In an IoT environment, the most common communication is M2M (machine-to-machine), so there should be trust between two machines before they share information. For proper functioning of IoT system, trust between each layer (perception layer, network layer, processing layer, and application layer) of IoT architecture is also important. So, there should be proper communication between each layer while keeping the data secure—data cannot be manipulated in between the communication.

As IoT is a conjunction of various other technologies such as cloud computing, RFID, sensor network, mobile communication, and so on, the security threat in all these technologies are, by default, inherited to IoT systems [17]. The most important objective in IoT security is to secure the data as it can contain sensitive information.

As the use of IoT in various sectors like the smart house, smart city, agriculture, home automation, industry, etc. is growing at a faster rate, the security issues need to be addressed quickly. For example, on December 23, 2015, in a part of Ukraine, cyber attackers cause the first blackout. This is the first-ever publicly acknowledged cyberattack on the power grid. The hackers attacked the information system of the three power distribution industry in Ukraine that caused a power outage. This incident affected about 2,25,000 customers and lasted for several hours [19]. In the next example, we will explain how attackers can cause a potential threat to a smart house. Adversaries can attack smart house system in several ways—attackers can install a rogue app on the house owner's phone and can control devices inside the house and switch on the light while house owner is sleeping; they can also take data from various IoT devices installed in that house to figure out if the house owner is present or when the house is likely to be empty [18].

From the two examples, we see how adversaries can crash the IoT system, causing sensitive data leakage. These IoT devices are easy targets of attackers due to poor security practices and heterogeneity. As the devices are for specific task operation, adding security solution to safeguard the data may affect device performance, power consumption, consumer dissatisfaction as well as the device performance.

To secure IoT, devices should have the ability to authenticate any entity that is trying to get access to the device; software in all devices also need to be authorized [20]. In IoT, data security is significant as the authorized user may have the chance of losing some sensitive information. For data security in IoT, a system should consider five basic security characteristics—data confidentiality, data integrity, data access control, data availability, and data encryption.

5.1 Data Confidentiality

Data confidentiality is about keeping the data secret from third party access and protect it from an unauthorized entity (person or thing) [18]. While sharing information from the sensor node to cloud or internet, the neighbor node cannot gather data [21]. This is the most important security principle in industrial, military, healthcare, and social application of IoT. For example, in health care the basic principle of data confidentiality is that the patient data should be provided to the patient as well as the doctors authorized to access that data and should be prohibited to any other staff who do not provide care to the patient.

5.2 Date Integrity

This means the data should not be modified or corrupted over its entire life cycle. Sometimes, the adversaries try to corrupt the data when they realize that it is impossible to access the data from the device. So, IoT device should ensure that the data is not altered, destroyed, or corrupted by any entity accessing that data. It should also ensure that the communication channel is reliable (i.e. while data transmission take place from one layer the next the data is kept original).

5.3 Data Access Control

For secure communication of two or more devices, trust is a significant factor. As emphasized in [20], the most significant factor while accessing data on any other entity is to identify the most appropriate owner of the data based on how sensitive the information is. There are various authentication mechanisms like biometrics authentication, cryptographic authentication, password-based authentication and so on. But this authentication mechanism does not work effectively in an IoT environment because some devices have low memory and processing power. A strong and lightweight authentication system serves well in this heterogeneous network.

5.4 Data Availability

The vision in IoT is to enhance the ability of a device to connect with the user and any other entity when needed. The data should be easily available to the user anytime and anywhere. We should not think that data availability means giving access to any unauthorized entity. Data availability is crucial in the field of IoT as low availability of data can impact any organization financially and make it unable to perform and provide services to clients. For example, suppose a person travelling on a train is trying to access the company project but due to limited data access, he is unable to work on it. Data availability is also important in the healthcare system, doctors cannot provide proper care to the patient if they have limited access to patient data. So, the authorized user should be able to access the data even in adverse conditions.

5.5 Data Encryption

In IoT, there should be a provision of end-to-end encryption so that there will not be any leakage of data. In an encryption mechanism, the data is first changed from plaintext to ciphertext, and then decrypted to get the actual data.

The cryptographic algorithms currently in use are of two types: symmetric cryptosystem and asymmetric cryptosystem. In symmetric cryptosystem, the receiver and the sender both use a similar key for encryption and decryption purpose. This symmetric cryptographic algorithm uses a secure channel to transfer the secret key; both parties should get access to the secret key. The drawback of this cryptographic algorithm is that with the increase in the number of users, it is impossible to keep the key secret. Some of the examples of symmetric cryptosystems are Blowfish, AES, DES, etc. In asymmetric cryptosystem, data is encrypted with the help of public key and then the ciphertext is decrypted by the receiver with the help of private key (this key is different from the public key and could not be obtained by remodeling the public key). Diffie and Hellmann proposed the first asymmetric cryptosystem. Some examples of these cryptosystems are RSA and ECC.

This encryption techniques ensure the authenticity and confidentiality of the data. As IoT is a heterogeneous environment, some devices have limited resources like low processing power, low battery life, small memory, etc. For these low-resource devices, the conventional cryptosystem do not fit well. Let's say, in RIFD tag, the most common cryptosystem 1204-bit—RSA algorithm do not work [22].

As explained in [23], these strong cryptosystems provide data security to devices with limited resources and cope with the trade-off between security, cost, and performance. For example, TEA (Tiny Encryption Algorithm) based on Feistel type cipher works well with these low-resource devices and provide strong security[24].

6 LAYERED ANALYSIS OF ATTACKS AND COUNTERMEASURES

In this section, we will explain why securing each layer is necessary with the help of various attacks and their countermeasures. As explained in the previous section, the security of IoT is necessary to safeguard the devices and the data communication path.

Security in the IoT environment is explained as a process of shielding physical resources and sensitive information from adverse attacks like hardware damage, fake node injection, or sinkhole attacks. As IoT is growing rapidly, the introduction of new limited-resourced devices (explained in section III) leads to new security attacks. Considering its faster growth, it is necessary to shield these devices with appropriate countermeasures.

Before going for various attacks that can be performed by the adversary, this paragraph will give a brief definition that what an attack is. As explained in [25], attacks are defined as any action taken to damage a system to gain unauthorized access or interrupt the normal functionality of the system by exploiting the vulnerabilities (here. it refers to the weakness or defects in a system) of that specific system. The attack cost is defined by the effort that the attacker put measured in terms of expertise, resource, and motivation [17,25].

We will discuss some of the major attacks taxonomy like active attack, passive attack, logical attack, and physical attack that we will use while analyzing the types of attack in each layer. An active attack is an attack where the adversary tries to modify or damage the data or the devices in the IoT environment. This attack causes damage to the system and breaks the data transmission path. The passive attack is completely opposite; in this attack, the adversary poses a threat to the confidentiality of data by eavesdropping or monitoring the data communication path. This attack poses no damage to any resource in the IoT platform [26,27].

In logical attack, the adversary does not cause any damage to IoT components but attack the data communication channel; in physical attack, the attacker damage or modify the infrastructure of IoT physically.

6.1 Perception Layer Attacks and Countermeasures

6.1.1 Some Attacks on Perception Layer

a Fake Node Insertion

In this type of attack, the attacker can inject a fake node between the actual nodes of a wireless network to access the IoT environment [28]. Further, they can direct the flow of data and use the IoT network for malicious purposes. If it is not prevented, it can gradually destroy the whole IoT network [29].

b Malicious Code Insertion

This is a type of physical attack where the attacker injects malicious code into the device via software/hardware interfaces present on it. Malicious code can be in many form like viruses, worms, trojan horses, backdoors and harmful content. This attack causes loss of precious resources of the network and causes a burden on the functionality of physical devices.

c Side Channel Attack

In this attack, the attacker tries to get into the network using physical features of IoT devices like power dissipation, electromagnetic emanation, running time, temperature, and acoustics. Some of the classes of side channel attack include cache attack, timing attack, power-monitor attack, electromagnetic attack, acoustic cryptanalysis, and differential fault analysis [30]. Power monitor attack exploits the data of a device's power consumption that depends on parameters like data processed and the number of instructions implemented, while electromagnetic attack can exploit the data of device's electromagnetic emanations.

d Sinkhole Attack

This is a type of intrusion attack in which adversary hacks a node or inserts a malicious node that presents itself as the optimum path to the base station. It then diverts the traffic generated by the other nodes towards itself by using false routing information. The counterfeit node generally attracts the nodes closer to the base station. It can also modify the data flowing towards the base station, thus compromising the security of IoT network. In [31], the authors have compiled a survey that analyzes different approaches of sinkhole attack detection like rule based detection, anomaly based detection, detection using statistical method, and hybrid-based intrusion detection.

e Device Tampering

This attack aims at damaging and exploiting the devices physically in an IoT environment. This attack can be considered as a hardware-centric attack that affects the nodes and interrupts its normal operation.

This attack itself is responsible for or acts as a gateway for various other attacks like DoS attacks, fake node injection attack, etc. By disrupting the physical network node operation, the attacker can even get sensitive information like a cryptographic key, radio key, etc. and poses a threat to all the higher layers in IoT architecture [3,32].

f Social Engineering Attack

Most of us think that cryptosystem changes the data to ciphertext before communicating it, or security principles like access control system make sure that no attacks can be performed on the device but, in this type of attack, no security principle works as the adversary emotionally, physically, or in both ways attack the authorized user for getting information without using any adversarial hacking device. These are the types of attacks in which the manipulator manipulates the knowledge workers, any organization physically, or psychologically to get access to critical information. The knowledge workers refer to those workers whose main capital is knowledge or information [33,34].

Reverse Social Engineering is the same as social engineering but in this, the attackers use a different approach—the attacker manipulates the authorized user by making him believe that he is a trustworthy entity and gets access to the data or information [35].

g Node Capture Attacks

This attack is most common in WSN (Wireless Sensor Network). In this attack, the intruder performs active, passive, and hardware attack collectively. Due to the huge application of WSN in different sectors like healthcare and military, it is necessary to address its threats. As previously stated, WSN consists of densely deployed sensor nodes over a sensor field. These sensor nodes are limited-resource devices (i.e. these work on limited battery, low memory, and low processing power) and provide hop to hop network connectivity so these are prone to security attacks like eavesdropping and node capture attack [36].

In this attack, while targeting a sensor network device the attacker tries to extract information through the data transmission channel by eavesdropping or constant monitoring. Then, the attacker uses various resources to access the information and, in case the data payloads in the transmission unit are encrypted, the adversary tries to figure out the network structure by learning the network operation. After various active and passive learning of the network or the data communication path, the intruder captures the sensor node based upon the task he needs to perform. To perform this attack, the adversary has to use various adversarial devices, so the attack cost of this is higher in comparison to other attacks [36,37].

h Sleep Deprivation Attack (SDA)

SDA deprivation attack, also known as resource consumption attack, is an attack where the adversarial device or node try to maximize the power consumption of the victim device or node to consume the battery [38]. This type of attack is common in constrained devices that have limited battery life. Because of limited resources in these devices, shielding them with strong security mechanism is difficult. Detecting such an attack is difficult so the only way is to use a proper security arrangement that alerts in advance before any device is turned off.

Let us see how this attack is done by taking the example of Wireless Sensor Network (WSN). In WSN, the sensor node used is resource-limited devices with limited battery life and limited processing power. As these sensor nodes have limited battery, they cannot operate for a long period without going into sleep mode. Therefore, researchers are motivated to use the sensor node clustering technique that interconnects various sensor nodes and partition the nodes into different groups, where a single node act as a cluster head. This cluster head collects data from participating nodes and directs them to database. This clustering gives an advantage to the adversary to successfully launch this SDA attack. As shown in Figure 9.5, the adversary can disturb the clustering algorithm and declare the adversarial node as the cluster head, or the adversarial device can be selected automatically to become a cluster head. Then, it keeps interacting with the victim cluster node by forwarding unnecessary packages to minimize its battery [39,40].

6.1.2 Some Countermeasures to Protect Perception Layer

a Authentication

As we previously explained, authenticity is necessary in IoT as it ensures that no unauthorized user or malicious device can take part in IoT network and get access to the data. Without a proper authentication mechanism, the adversary can hack the system and disrupt the normal operation of component in IoT environment. While implementing the authentication mechanism, we should ensure that it makes the device secure as well as transient. Here, transient means that when the authorized user of the device is changed the device should work in accordance to the current user and should not leak data of its previous user [41].

There are various authentication techniques like biometric authentication, hybrid text-based authentication, multi layer authentication, and password-based authentication. We will explain some of the proposed methods by different researchers. In [42], a client-based user authentication agent technique is proposed, where the user has to download an allowance with unique access code from the service provider and the server encrypts the code and the client decrypts it with a unique password. This mechanism is based on SaaS (Software-as-a-Service) with enhanced Diffie-Hellman algorithm [43]. In [44], the authors have proposed a lightweight

Analysis of Security Issues in IoT System

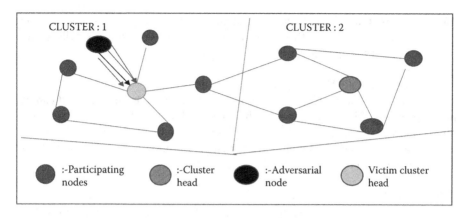

FIGURE 9.5 Sleep Deprivation Attack.

authentication protocol using XOR operation that provides a strong authentication mechanism to resource constrained devices. In [45], the authors provide a multitier authentication mechanism comprised of two tiers—the first tier use simple password authentication mechanism and, in the second tier, the user enters a predetermined sequence while the authorized entity is watching it.

b Data Integrity Schemes

The data integrity scheme makes sure that the data is not manipulated or corrupted by the unauthorized user during the process of data transmission. These schemes prove verification of the data generated by a device as well as the data stored in the end database [41,46].

In [47], the author proposed a method to preserve data integrity by using a Bayesian inference model. This model measures the reliability of data and categorize them into not compromised, compromised, and cannot be inferred. [46] purposed a method to hide the data before data transmission. This proposed method uses a random time hopping sequence to conceal the data. In this method, a secret seed is shared between the server and the device, then the secret seed is used to generate the RTH.

c IPSec (Internet Protocol Security)

This security protocol suit provides three security features (i.e. data confidentiality, data authentication, and integrity verification). This is mostly used in the network layer and proved to be the most effective countermeasure against replay attack, eavesdropping, and authentication attack. This can work with several cryptographic algorithms to encrypt the data before package transmission. The most common algorithm used is a triple data encryption algorithm (TDEA) that provides data confidentiality and hash-based message authentication code (HMAC) which verify the authenticity of data [48,49].

This protocol suit has two modes (i.e. transport mode and tunnel mode.) Transport mode verifies the IP payload only, while tunnel mode encapsulates the whole IP datagram with tunneled IP projection. This protocol suit is further divided into three sub-protocols (i.e. Encapsulating Security Protocol (ESP), Authentication Header Protocol (AHP), and Key Exchange Protocol) where each work collectively to prove IPSec protection [49,50].

d Secure Physical Designing

A large portion of the perception layer consists of hardware devices. Thus, it can be made secure by efficient designing of the physical devices. The components of the physical devices should be made of high quality material so that they are not damaged easily. The devices should have robust composition so that they last longer in the open environment.

e Safe Booting

Booting is defined as powering up devices to make them ready for operations. The perception layer can be made safe if we employ secure booting techniques. In secure booting, the operating system of device utilizes cryptographic algorithms to ensure the integrity of devices by checking for malware and protecting against malicious code and unauthorized software updates.

6.2 Network Layer Attacks and Countermeasures

6.2.1 Some Attacks on Network Layer

a Traffic Analysis Attack

In this attack, the attacker intercepts the flow of data between communications and tries to analyze the data to gather confidential information for their malicious intent. Traffic analysis attack even works on encrypted data. Various parameters like the volume of data, data frequency, data size, and data pattern are of particular interest to the attacker. The attacker takes the aid of sniffing programs like port scanning application and packet sniffer applications before deploying the attack [51]. As the network layer is the connective layer connecting between perception layer and other layers, this attack can potentially disrupt the entire IoT network.

b Man in the Middle Attack (MIMA)

In this type of attack, the adversary tries to intercept and eavesdrop on the communication between the two ends. Here, the attacker does not have to be physically present near the network, they can just exploit the communication protocols in the network layer to get classified information [52]. The attacker may steal confidential data or alter/delete data for their malicious intent. This is a type of active attack as the attacker has to persuade both ends that he is the other end. Then, the attacker has to also keep fluent communication between the two ends while impersonating as the other.

c Denial of Service Attack (DoS)

In this attack, the adversary tries to render the network services inaccessible to a user. The attacker can either do application level flooding attacks that exploit the user's services by draining the server resources, or a network level flooding attack that severs the user's connectivity by draining the network resources. Thus, a DoS attack affects the computational power of the network and disrupts the flow of data. There are two types of DoS attacks: simple DoS attack and distributed dos attack (DDoS).

d Sybil Attack

This is the most dangerous attack on a peer-to-peer system like WSN and ad-hoc network. In this attack, the adversarial node can mimic multiple identities. These Sybil identities manipulate the reputation system by sending false message or warning signal. An adversary can use this technique to launch other attacks like DoS and routing attack [53,54].

e False Data Injection Attack (FDIA)

In the IoT environment, the network of nodes is mainly responsible for data transmission. These nodes are protected by various security algorithm and cryptosystem. In FDIA, the attacker tries to figure out a small error in these algorithms and inject false data without alerting security protocols. This attack causes false report and can also maximize power consumption of nodes by generating pseudo events similar to that of sleep denial attack [55].

f Black Hole Attack (BHA)

This security attack is common in the ad hoc network. The attacker, with the help of adversarial node, try to manipulate the victim node whose packets it wants to intercept and create a pseudo route. The adversary node tries to show the victim node that it is having the best and shortest route for sending network packets. With the help of this attack, the attacker can launch various other attacks like DoS, MIMA, and FDIA.

g Worm Hole Attack

This is a passive attack where the adversarial node tries to receive data packets from the victim node and send them to another through a wormhole tunnel. The wormhole tunnel takes advantage of weak routing protocol security, and can threat data integrity and confidentiality, as well as create a hotspot for several other attack [56]. As explained in [57], wormhole attack in dynamic source routing protocol and ad hoc on-demand distance vector protocol can even prevent the network system from finding routes other than the wormhole tunnel.

h Routing Table Overflow Attack

In this, the adversary advertise new routes to nonexistence node and confuse the distributed system by preventing it from creating routes.

i Distributed Denial-of-Service Attack

We already discussed about DoS attack in this section. DDoS attack is similar to DoS attack but the only difference is that, in this attack, the attacker deploys multiple adversarial nodes or entities to flood the bandwidth or resources of an entity, making the services inaccessible to the authentic user. This attack can be performed in various ways like the attacker can confuse the system by sending compromised packets, or the attacker can figure out security holes and take advantage of the device security vulnerabilities to infect the device with pseudo code and attack packets. In this attack, the attacker tries to hide the identity of the subverted entity by spoofing the source address [58,59].

6.2.2 Some Countermeasures to Protect Network Layer

a Routing Security

As mentioned in [60], some of the major principles that we should follow for safe routing are maximizing the network quality of service, decreasing the consumption, securing the high sensitive flow, and checking the difference between the data packets from both the sender and receiver side.

b Protection against Denial of Service Attacks

In [61], the authors have explained some methods to prevent DoS attacks—detection, classification, and response. Detection refers to analyzing deviation from normal behavior or identifying characteristics according to previous attacks. Classification analyzes the data packets used in communication and classifies it into normal and DoS packets. Response refers to the act of analyzing DoS packet and dropping them in the constant interval.

In [62], the research presents some approaches to protect the network against DDoS attacks—source-based, destination-based, network-based, and hybrid approaches. In a source-based approach, the protective measures are applied at the source end. In destination-based approach, the protective measures are applied at the destination end; in network-based approach, the protective measures secure the network medium. In hybrid approach the protective measures are applied at various strategic locations to efficiently secure the IoT network.

c Sybil Attack Countermeasure

[63] proposed a lightweight intrusion detection algorithm for Sybil attack with an accuracy of 95% in mobile RPL (Routing Protocol for Low Power and Lossy Network). The accuracy is calculated based on factors like package delivery ratio, energy consumption, and accuracy in detection. This detection mechanism works well with limited-resourced devices and has low computational complexity. In [64], the author proposed a way to enhance LEACH protocol security with the help of jakes channel model and RSSI (Received Signal Strength Indicator) detection mechanism for Sybil attack.

d False Data Injection Attack Countermeasures

[65] proposed a Kalman filter along with Euclidean detector in detecting FDIA. This Kalman filter, along with $\chi 2$-detector, can also be used to detect other attacks as well, like DoS. In [66], the author tried to figure out the effect of FDIA on CSSVC (Control Signal from the controller to the Static Var Compensator) and NVSI (Node Voltage Stability Index), and then propose a clustering algorithm that uses topology of the power system, CSSVC value and NVSI value, to find out that if any node exposed an FDIA attack. Then, this mechanism used a forecasting method to detect FDIA. [67] proposed two filtering mechanisms (i.e. GFFS (Geographical information based False data Filtering Scheme) and NFFS (Neighbor information based False data Filtering Scheme)). GFFS uses the absolute position of

Analysis of Security Issues in IoT System 163

the sensor, while NFFS use relative position based on the neighboring sensor node. In this scheme, each node sends a data report (containing message authentication code and location) to the forwarding nodes.

6.3 Processing Layer Attacks and Countermeasures

6.3.1 Processing Layer Attacks

a Session Hijacking

This is a type of attack where the adversary attempts to hijack the authorized user web session either by filching the session cookie or by guessing it. In today's web service, cookie is commonly used as user identification. This cookie is a token that acts as user identity, used by the service provider to provide personalized search as well as to know client interest. This type of attack poses a threat towards user data integrity and confidentiality [68].

b XML Signature Wrapping Attack

Digital signatures are mostly used to protect documents and sensitive messages. To create signature, the sender uses a signature algorithm that takes private key as input and the valid receiver can only get access to the document if he has a valid public key provided by the sender. In this attack, the adversary attempts to disrupt the signature algorithm and inject false data in the XML documents while keeping the original data intact [69].

c SaaS Security Threats

The security of client data is one of the biggest threat in SaaS (Software as a Service). In SaaS, the client depends on the service provider to ensure security and safeguard the data. This third party (service provider) access increases the threat to data confidentiality [70,71].

d SQL Injection

In this type of attack, the attacker tries to exploit the database of a network by using a malicious code. The malicious code is generally a well-crafted SQL command that causes the database to execute actions according to the attacker's intent.

e Flooding Attack

In this type of attack, the attacker tries to exhaust the servers by making large number of requests. This attack targets the network resources and, thus, reduces the quality of service provided.

6.3.2 Processing Layer Countermeasures

a Homomorphic Encryption

As we discussed earlier, in storing data to the cloud, there is a chance of third party access. To prevent this, the homomorphic encryption scheme is helpful [72]. In this encryption scheme, the cloud provider can do various computation on the data without decrypting it, while in normal encryption scheme the data needs to be decrypted before executing any computation on it.

b End-to-End Encryption

Encryption is necessary to secure the data; in no end-to-end encryption mechanism, during the data transmission from one protocol to another, there is a need to decrypt the data at gateways level. In the end-to-end encryption mechanism, only the client or the service provider can decrypt the data, making it much more secure [73].

6.4 Application Layer Attacks and Countermeasures

6.4.1 Some Attacks on Application Layer

a Data Modification Attack

In this attack, the attacker tries to modify the data that is being transmitted from sensors to the base station. By sending modified data, the attacker can manipulate the base station to make a distorted decision that is beneficial to the attacker instead.

b Reprogramming Attack

In this attack, the attacker attempts to modify the source code of the program for their malicious intent [74]. The attacker makes the application inaccessible to the user by forming an infinite loophole in the program.

c Phishing Attack

In this attack, the adversary tries to steal sensitive data like passwords and user IDs by acting as a trustworthy source. Sensitive data is generally accessed by using fake websites and spoofed emails [75]. The attacker can use the sensitive data for further disrupting the IoT network for their malicious purposes.

a. Sniffer Attack
   This is an attack where the adversary steals sensitive information of the authorized user with the help of sniffing tools and programs. The adversary uses these sniffing programs to capture data packages that contain information of the user, like user ID, passwords, and other sensitive data [73].
b. Access Control Attack

In this attack, the unauthorized user attempts to get access to the control system as well as the sensitive data of the user. Access control attacks can potentially cause huge financial loss to the user in different IoT application sectors like healthcare, industry, etc. For example, if an adversary gets unauthorized access to the control system of an industry, then he can cause devastating damage to devices as well as financial loss to the owner.

6.4.2 Some Countermeasures to Protect Application Layer

a Firewall

A firewall is a defense mechanism that protects a private network from unauthorized malware like virus, spam, and spyware from the internet.

Analysis of Security Issues in IoT System

Perception layer	Network layer	Processing layer	Application layer
Attacks • Fake node insertion • Malicious code insertion • Side channel attack • Sinkhole attack • Sleep deprivation attack • Social engineering attack • Device tampering • Node capture	**Attacks** • Traffic analysis attacks • Man in the middle attack • Denial of services attack • Sybil attack • Routing table overflow attacks • False data injection • Black hole attack • Wormhole attack	**Attacks** • Session Hijacking • Flooding attacks • SQL injection attack • XML Signature wrapping attack • SaaS security threats	**Attacks** • Data modification attack • Access control attack • Phishing attack • Reprogramming attack • Sniffing attack
Countermeasures • Secure physical designing • Safe booting • Authentication • Data integrity schemes	**Countermeasures** • Routing security • Protection against denial of service attacks • Sybil attack countermeasures • False data injection countermeasures	**Countermeasures** • End to end encryption • Homomorphic encryption	**Countermeasures** • Firewall • User authentication • Sniffing detection techniques

FIGURE 9.6 Table of All Attacks and Their Respective Counter Measures

b User Authentication

We previously discussed various authentication mechanism like data signature that protect the data from the attacker. In the application sector, authenticity schemes generally deal with securing devices from unauthorized access, so there should be an authentication mechanism to decide if an entity should be given access to the device or not.

c Sniffing Attack Countermeasure

Sniffing attack is one of the most dangerous attacks that targets authorized user sensitive data. [76] proposed two techniques to secure data packages from a sniffing attack. This proposed mechanism uses APR (Address Resolution Protocol) and RTT (Round Trip delay Time) to detect sniffer attack (Figure 9.6).

7 CONCLUSION

IoT is a revolutionary technology but it is still not fully integrated with our society. The main reason is due to its security vulnerabilities. IoT is a conjunction of different technologies, so it deals with the security issues of different technologies altogether. To further enhance the security aspect, we need to merge emerging technologies like blockchain, 5G network, and AI in IoT environment. Blockchain is a distributed database mostly used in bitcoin cryptocurrency. This technology is fast, trustworthy, and forms a secure mesh network that allows various entities to interact, avoiding a spoofing attack. The emerging of 5G technology eradicates the connectivity problems in an IoT environment. Similarly, AI and machine learning can analyze the attack pattern with the help of complex pattern recognition algorithm to provide better security.

In this book chapter, we discussed various technologies involved in IoT. Then, we discussed how the four layers of IoT work together. The attacks, along with their respective countermeasures, have also been discussed. This chapter is expected to give an insight into security vulnerabilities of the different layers along with their countermeasures, and can be served as an asset for the development of IoT.

REFERENCES

1. Kumar, N. M. and Mallick, P. K. (2018). The internet of things: Insights into the building blocks, component interactions, and architecture layers. *Procedia computer science*, *132*, pp. 109–117.
2. Bello, O., Zeadally, S., and Badra, M. (2017). Network layer inter-operation of device-to-device communication technologies in Internet of Things (IoT). *Ad Hoc Networks*, *57*, pp. 52–62.
3. Ahemd, M. M., Shah, M. A., and Wahid, A. (2017, April). IoT security: A layered approach for attacks & defenses. In 2017 International Conference on Communication Technologies (ComTech) (pp. 104–110). IEEE.
4. Jurcut, A. D., Ranaweera, P., and Xu, L. (2020). Introduction to IoT security. *IoT security: Advances in authentication* (pp. 27–64). Wiley.
5. Raza, S., Duquennoy, S., Höglund, J., Roedig, U., and Voigt, T. (2014). Secure communication for the Internet of Things—a comparison of link-layer security and IPsec for 6LoWPAN. *Security and Communication Networks*, *7*(12), pp. 2654–2668.
6. Sun, X. and Ansari, N. (2017). Dynamic resource caching in the IoT application layer for smart cities. *IEEE Internet of Things Journal*, *5*(2), pp. 606–613
7. Jia, X., Feng, Q., and Ma, C. (2010). An efficient anti-collision protocol for RFID tagidentification. *IEEE Communications Letters*, *14*(11), pp. 1014–1016.
8. Peris-Lopez, P., Hernandez-Castro, J. C., Estevez-Tapiador, J. M., and Ribagorda, A. (2006, September). RFID systems: A survey on security threats and proposed solutions. In IFIP International Conference on Personal Wireless Communications (pp. 159–170). Springer, Berlin, Heidelberg.
9. Guo, L. G., Huang, Y. R., Cai, J., and Qu, L. G. (2011, July). Investigation of architecture, key technology and application strategy for the Internet of Things. In Proceedings of 2011 Cross Strait Quad-Regional Radio Science and Wireless Technology Conference (Vol. 2, pp. 1196–1199). IEEE.
10. Gubbi, J., Buyya, R., Marusic, S., and Palaniswami, M. (2013). Internet of Things (IoT): A vision, architectural elements, and future directions. *Future Generation Computer Systems*, *29*(7), pp. 1645–1660.
11. Islam, S. R., Kwak, D., Kabir, M. H., Hossain, M., and Kwak, K. S. (2015). The internet of things for health care: A comprehensive survey. *IEEE Access*, *3*, pp. 678–708.
12. Gaitan, N. C., Gaitan, V. G., and Ungurean, I. (2015). A survey on the internet of things software architecture. *International Journal of Advanced Computer Science and Applications (IJACSA)*, *6*, p. 12.
13. Bhoi, S. K., Panda, S. K., Padhi, B. N., Swain, M. K., Hembram, B., Mishra, D., ... Khilar, P. M. (2018, December). Fireds-iot: A fire detection system for smart home based on iot data analytics. In 2018 International Conference on Information Technology (ICIT) (pp. 161–165). IEEE.
14. Bhoi, S. K., Panda, S. K., Patra, B., Pradhan, B., Priyadarshinee, P., Tripathy, S., ... Khilar, P. M. (2018, December). FallDS-IoT: A fall detection system for elderly healthcare based on IoT data analytics. In 2018 International Conference on Information Technology (ICIT) (pp. 155–160). IEEE.
15. Abuarqoub, A., Abusaimeh, H., Hammoudeh, M., Uliyan, D., Abu-Hashem, M. A., Murad, S., ... Al-Fayez, F. (2017, July). A survey on internet of things enabled smart campus applications. In Proceedings of the International Conference on Future Networks and Distributed Systems (pp. 1–7).
16. Srivastava, P., and Garg, N. (2015, May). Secure and optimized data storage for IoT through cloud framework. In International Conference on Computing, Communication & Automation (pp. 720–723). IEEE.

17. Abomhara, M. (2015). Cyber security and the internet of things: Vulnerabilities, threats, intruders and attacks. *Journal of Cyber Security and Mobility*, 4(1), pp. 65–88.
18. Brauchli, A. and Li, D. (2015, August). A solution based analysis of attack vectors on smart home systems. In 2015 International Conference on Cyber Security of Smart Cities, Industrial Control System and Communications (SSIC) (pp. 1–6). IEEE.
19. Case, D. U. (2016). Analysis of the cyber attack on the Ukrainian power grid. *Electricity Information Sharing and Analysis Center (E-ISAC)*, Vol. 388.
20. Mahmoud, R., Yousuf, T., Aloul, F., and Zualkernan, I. (2015, December). Internet of things (IoT) security: Current status, challenges and prospective measures. In 2015 10th International Conference for Internet Technology and Secured Transactions (ICITST) (pp. 336–341). IEEE.
21. Mishra, S. K., Sahoo, B., and Jena, S. K. (2019). A secure VM consolidation in cloud using learning automata. In Recent Findings in Intelligent Computing Techniques (pp. 617–623). Springer, Singapore.
22. Singh, S., Sharma, P. K., Moon, S. Y., and Park, J. H. (2017). Advanced lightweight encryption algorithms for IoT devices: survey, challenges and solutions. *Journal of Ambient Intelligence and Humanized Computing*, pp. 1–18.
23. Eisenbarth, T., Kumar, S., Paar, C., Poschmann, A., and Uhsadel, L. (2007). A survey of lightweight-cryptography implementations. *IEEE Design & Test of Computers*, 24(6), pp. 522–533.
24. Andem, V. R. (2003). *A Cryptanalysis of the Tiny Encryption Algorithm*. Doctoral dissertation, University of Alabama.
25. Bertino, E., Martino, L. D., Paci, F., and Squicciarini, A. C. (2009). Web services threats, vulnerabilities, and countermeasures. In *Security for web services and service-oriented architectures* (pp. 25–44). Springer, Berlin, Heidelberg.
26. Nawir, M., Amir, A., Yaakob, N., and Lynn, O. B. (2016, August). Internet of Things (IoT): Taxonomy of security attacks. In 2016 3rd International Conference on Electronic Design (ICED) (pp. 321–326). IEEE.
27. Hossain, M. M., Fotouhi, M., and Hasan, R. (2015, June). Towards an analysis of security issues, challenges, and open problems in the internet of things. In 2015 IEEE World Congress on Services (pp. 21–28). IEEE.
28. Nia, A. M., and Jha, N. K. (2016). A comprehensive study of security of internet-of-things. *IEEE Transactions on Emerging Topics in Computing*, 5(4), pp. 1–19.
29. Ahemd, M. M., Shah, M. A., and Wahid, A. (2017, April). IoT security: A layered approach for attacks & defenses. In 2017 International Conference on Communication Technologies (ComTech) (pp. 104–110). IEEE.
30. Spreitzer, R., Moonsamy, V., Korak, T., and Mangard, S. (2017). Systematic classification of side-channel attacks: A case study for mobile devices. *IEEE Communications Surveys & Tutorials*, 20(1), pp. 465–488.
31. Kibirige, G. W. and Sanga, C. (2015). A survey on detection of sinkhole attack in wireless sensor network. *arXiv preprint arXiv:1505.01941*.
32. Deogirikar, J. and Vidhate, A. (2017, February). Security attacks in IoT: A survey. In 2017 International Conference on I-SMAC (IoT in Social, Mobile, Analytics and Cloud)(I-SMAC) (pp. 32–37). IEEE.
33. Krombholz, K., Hobel, H., Huber, M., and Weippl, E. (2015). Advanced social engineering attacks. *Journal of Information Security and Applications*, 22, pp. 113–122.
34. Krombholz, K., Hobel, H., Huber, M., and Weippl, E. (2013, November). Social engineering attacks on the knowledge worker. In Proceedings of the 6th International Conference on Security of Information and Networks (pp. 28–35).
35. Orgill, G. L., Romney, G. W., Bailey, M. G., and Orgill, P. M. (2004, October). The urgency for effective user privacy-education to counter social engineering attacks on

secure computer systems. In Proceedings of the 5th Conference on Information Technology Education (pp. 177–181).
36. Tague, P. and Poovendran, R. (2007). Modeling adaptive node capture attacks in multi-hop wireless networks. *Ad Hoc Networks*, 5(6), pp. 801–814.
37. Bonaci, T., Bushnell, L., and Poovendran, R. (2010, December). Node capture attacks in wireless sensor networks: A system theoretic approach. In 49th IEEE Conference on Decision and Control (CDC) (pp. 6765–6772). IEEE.
38. Bhattasali, T., Chaki, R., and Sanyal, S. (2012). Sleep deprivation attack detection in wireless sensor network. *arXiv preprint arXiv:1203.0231*.
39. Pirretti, M., Zhu, S., Vijaykrishnan, N., McDaniel, P., Kandemir, M., and Brooks, R. (2006). The sleep deprivation attack in sensor networks: Analysis and methods of defense. *International Journal of Distributed Sensor Networks*, 2(3), pp. 267–287.
40. Medeiros, H. and Park, J. (2009). Cluster-based object tracking by wireless camera networks. Elsevier.
41. Stajano, F. and Anderson, R. (1999, April). The resurrecting duckling: Security issues for ad-hoc wireless networks. In *International workshop on security protocols* (pp. 172–182). Springer, Berlin, Heidelberg.
42. Moghaddam, F. F., Moghaddam, S. G., Rouzbeh, S., Araghi, S. K., Alibeigi, N. M., and Varnosfaderani, S. D. (2014, April). A scalable and efficient user authentication scheme for cloud computing environments. In 2014 IEEE Region 10 Symposium (pp. 508–513). IEEE.
43. Saadeh, M., Sleit, A., Qatawneh, M., and Almobaideen, W. (2016, August). Authentication techniques for the internet of things: A survey. In 2016 Cybersecurity and Cyberforensics Conference (CCC) (pp. 28–34). IEEE.
44. Lee, J. Y., Lin, W. C., and Huang, Y. H. (2014, May). A lightweight authentication protocol for internet of things. In 2014 International Symposium on Next-Generation Electronics (ISNE) (pp. 1–2). IEEE.
45. Singh, A. and Chatterjee, K. (2015, March). A secure multi-tier authentication scheme in cloud computing environment. In 2015 International Conference on Circuits, Power and Computing Technologies [ICCPCT-2015] (pp. 1–7). IEEE.
46. Aman, M. N., Sikdar, B., Chua, K. C., and Ali, A. (2018). Low power data integrity in IoT systems. *IEEE Internet of Things Journal*, 5(4), pp. 3102–3113.
47. Bhattacharjee, S., Salimitari, M., Chatterjee, M., Kwiat, K., and Kamhoua, C. (2017, November). Preserving data integrity in iot networks under opportunistic data manipulation. In 2017 IEEE 15th International Conference on Dependable, Autonomic and Secure Computing, 15th International Conference on Pervasive Intelligence and Computing, 3rd International Conference on Big Data Intelligence and Computing and Cyber Science and Technology Congress (DASC/PiCom/DataCom/CyberSciTech) (pp. 446–453). IEEE.
48. Li, Z., Cui, X., and Chen, L. (2006, November). Analysis and classification of ipsec security policy conflicts. In 2006 Japan-China Joint Workshop on Frontier of Computer Science and Technology (pp. 83–88). IEEE.
49. Elkeelany, O., Matalgah, M. M., Sheikh, K. P., Thaker, M., Chaudhry, G., Medhi, D., and Qaddour, J. (2002, April). Performance analysis of IPSec protocol: Encryption and authentication. In 2002 IEEE International Conference on Communications. Conference Proceedings. ICC 2002 (Cat. No. 02CH37333) (Vol. 2, pp. 1164–1168). IEEE.
50. Guo, J., Gu, C., Chen, X., and Wei, F. (2019). Model learning and model checking of IPSec implementations for internet of things. *IEEE Access*, 7, pp. 171322–171332.
51. Thakur, B. S. and Chaudhary, S. (2013). Content sniffing attack detection in client and server side: A survey. *International Journal of Advanced Computer Research*, 3(2), pp. 7.

52. Rao, T. A. (2018). Security challenges facing IoT layers and its protective measures. *International Journal of Computer Applications*, 975, p. 8887.
53. Olakanmi, O. O., and Dada, A. (2020). Wireless Sensor Networks (WSNs): Security and privacy issues and solutions. In *Wireless mesh networks-security, architectures and protocols*. IntechOpen.
54. Douceur, J. R. (2002, March). The sybil attack. In International Workshop on Peer-to-Peer Systems (pp. 251–260). Springer, Berlin, Heidelberg.
55. Bostami, B., Ahmed, M., and Choudhury, S. (2019). False data injection attacks in internet of things. In *Performability in internet of things* (pp. 47–58). Springer, Cham.
56. Hu, Y. C., Perrig, A., and Johnson, D. B. (2006). Wormhole attacks in wireless networks. *IEEE Journal on Selected Areas in Communications*, 24(2), pp. 370–380.
57. Gagandeep, A., and Kumar, P. (2012). Analysis of different security attacks in MANETs on protocol stack A-review. *International Journal of Engineering and Advanced Technology (IJEAT)*, 1(5), pp. 269–275.
58. Mirkovic, J. and Reiher, P. (2004). A taxonomy of DDoS attack and DDoS defense mechanisms. *ACM SIGCOMM Computer Communication Review*, 34(2), pp. 39–53.
59. Feinstein, L., Schnackenberg, D., Balupari, R., and Kindred, D. (2003, April). Statistical approaches to DDoS attack detection and response. In Proceedings DARPA Information Survivability Conference and Exposition (Vol. 1, pp. 303–314). IEEE.
60. Papachristou, K., Theodorou, T., Papadopoulos, S., Protogerou, A., Drosou, A., and Tzovaras, D. (2019, June). Runtime and routing security policy verification for enhanced quality of service of IoT networks. In *2019 Global IoT Summit (GIoTS)* (pp. 1–6). IEEE.
61. Loukas, G. and Öke, G. (2010). Protection against denial of service attacks: A survey. *The Computer Journal*, 53(7), pp. 1020–1037.
62. Zargar, S. T., Joshi, J., and Tipper, D. (2013). A survey of defense mechanisms against distributed denial of service (DDoS) flooding attacks. *IEEE Communications Surveys & Tutorials*, 15(4), pp. 2046–2069.
63. Murali, S. and Jamalipour, A. (2019). A lightweight intrusion detection for sybil attack under mobile RPL in the internet of things. *IEEE Internet of Things Journal*, 7(1), pp. 379–388.
64. Chen, S., Yang, G., and Chen, S. (2010, April). A security routing mechanism against Sybil attack for wireless sensor networks. In 2010 International Conference on Communications and Mobile Computing (Vol. 1, pp. 142–146). IEEE.
65. Manandhar, K., Cao, X., Hu, F., and Liu, Y. (2014). Detection of faults and attacks including false data injection attack in smart grid using Kalman filter. *IEEE Transactions on Control of Network Systems*, 1(4), pp. 370–379.
66. Xu, R., Wang, R., Guan, Z., Wu, L., Wu, J., and Du, X. (2017). Achieving efficient detection against false data injection attacks in smart grid. *IEEE Access*, 5, pp. 13787–13798.
67. Wang, J., Liu, Z., Zhang, S., and Zhang, X. (2014). Defending collaborative false data injection attacks in wireless sensor networks. *Information Sciences*, 254, pp. 39–53.
68. Jain, V., Sahu, D. R., and Tomar, D. S. (2015). Session hijacking: Threat analysis and countermeasures. In International Conference on Futuristic Trends in Computational Analysis and Knowledge Management.
69. Kumar, J., Rajendran, B., Bindhumadhava, B. S., and Babu, N. S. C. (2017, November). XML wrapping attack mitigation using positional token. In 2017 International Conference on Public Key Infrastructure and its Applications (PKIA) (pp. 36–42). IEEE.
70. Puthal, D., Sahoo, B. P., Mishra, S., and Swain, S. (2015, January). Cloud computing features, issues, and challenges: A big picture. In 2015 International Conference on Computational Intelligence and Networks (pp. 116–123). IEEE.

71. Mishra, S. K., Puthal, D., Sahoo, B., Sharma, S., Xue, Z., and Zomaya, A. Y. (2018). Energy-efficient deployment of edge dataenters for mobile clouds in sustainable IoT. *IEEE Access*, *6*, pp. 56587–56597.
72. Mishra, S. K., Sahoo, S., Sahoo, B. (2019). Secure big data computing in cloud: An overview. *Encyclopedia of big data technologies 2019*. Springer.
73. Hassija, V., Chamola, V., Saxena, V., Jain, D., Goyal, P., and Sikdar, B. (2019). A survey on IoT security: application areas, security threats, and solution architectures. *IEEE Access*, *7*, pp. 82721–82743.
74. Sonar, K., and Upadhyay, H. (2014). A survey: DDOS attack on internet of things. *International Journal of Engineering Research and Development*, *10*(11), pp. 58–63.
75. Tewari, A., Jain, A. K., and Gupta, B. B. (2016). Recent survey of various defense mechanisms against phishing attacks. *Journal of Information Privacy and Security*, *12*(1), pp. 3–13.
76. Kulshrestha, A. and Dubey, S. K. (2014). A literature review on sniffing attacks in computernetwork. *International Journal of Advanced Engineering Research and Science (IJAERS)*, *1*(2).

10 Resource Optimization of Cloud Services with Bi-layered Blockchain

J. Chandra Priya and Sathia Bhama Ponsy R. K.
Department of Computer Technology, Anna University, MIT Campus, Chennai, India

1 INTRODUCTION

Cloud computing serves the on-demand provision of voluminous storage and heavy computing services dynamically over the internet. Cloud services offer a facility to hold, secure, and access the data on-the-go with a considerable response time with authorized resource escalation, whenever needed. Figure 10.1 elaborates on the range of services provided by cloud platforms. It is prone to failures and threats and avoids unauthorized access as presented by Rajathi et al. [1–4]. It renders the rented access of resources, hiding the irrelevant aspects such as infrastructure and locality of service. The cloud-moved data and applications are handled by several data centers that are not fully distributed in the present cloud storage [5–7]. The data are dumped in various data centers at high volume, and there are feasibilities for data to be lost if any of the data centers were damaged. This data migration transfers the administrative control from the data owner to the third-party service providers. But there is still no practical solution for the security of distributed cloud storage, as criticized by Tosh et al. [8]. Several cryptographic solutions have been proposed [9] to overcome the issues mentioned earlier, which lags in the implicit vulnerabilities and key management.

Blockchain is an emerging technology that imitates the renowned linked list data structure that holds transactions as blocks to be cryptographically back-linked chains [10]. It is a distributed ledger in a decentralized environment that utilizes game theory to achieve an agreed-upon rule to add the block to the chain. Asymmetric key cryptography is utilized in such a way that secret keys are used to generate a digital signature for a transaction, and public keys to verify the signature. Public keys generate pseudonymous addresses by hashing [11]. A digital wallet can be viewed as a key store in this context [12,13]. Blockchain is used to store data from multiple systems into a decentralized ledger and forms a more trusted environment that manages the ledger [14]. Entire transaction data can be queried, and a transparent log of information can be obtained. Smart contract codes can be run on top of blockchain to enforce rules, as mentioned by Christidis

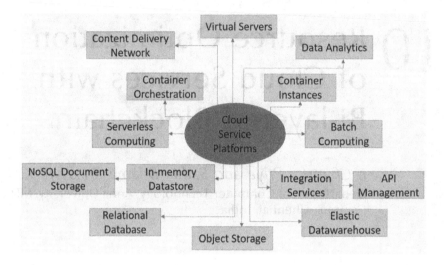

FIGURE 10.1 Range of Services Provided by Cloud Service Providers.

et al. [15] and Heilman et al. [16]. In case of issues, there are provisions to trace the root cause of defects quickly [12] and take measures on the defect, as shown in Figure 10.2.

This work deals with the need for blockchain in cloud service platforms to make cloud storage reliable, faster, and more secure than other storage and computing technologies.

2 LITERATURE REVIEW

Bogumil et al. [17] proposed a simulator for Elastic Cloud Computing spot pricing mechanisms. This simulator has a provision to develop and test the bidding strategies on the Amazon spot price market. The main goal of this work is to reduce the

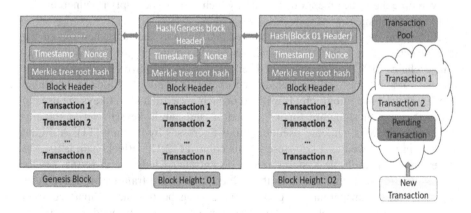

FIGURE 10.2 Skeleton of a Blockchain.

computational cost and time of running computing clusters with spot pricing mechanisms. But their bidding strategies become invalid after extended periods. Al Omar et al. [18] proposed a system for privacy-preserving platforms for healthcare data in the cloud. Cloud stocked data has ever been vulnerable to hacks. So, they used decentralization for those cloud data that can minimize the effect of attacks. Several methods and techniques prevail to control the impact of cyber-attacks, especially with decentralized approaches [19]. The performance evaluation shows that this platform runs well in the blockchain environment but, in another direction, does not address the issues of handling key-theft/loss mechanisms or key distribution techniques.

Z. Zhang et al. [20] proposed that the file creator has given them permission to supply the secret key for the requestors and encipher the requested data by assigning access policies, and the proposed scheme attains a fine-grained access check on data. Concurrently, the smart contract written on top of the Ethereum blockchain [21] implemented a keyword search function on the ciphertext of the decentralized storage systems so that the traditional ABE solution always requires the existence of a trusted private key generator (PKG), as criticized by Zyskind et al. [22,23]. The private key generated by the PKG for the users is not flexible enough and may result in key abuse. Zyskind et al. [22] proposed a controllable and trustworthy blockchain-based cloud data management system that covers tracing and auditing the unauthorized modification that happened in a broad range of applications by using the Controllable Block Chain Data Management (CBDM) model that can be set out in the cloud environment. However, the approach is not simulated in real-world environments, and they missed covering the part of the approach for majority attacks.

Kecong et al. (2018) [9] provided a solution for storing usernames and passwords of people using blockchain in an encrypted format. As a result, people do not need to memorize their usernames and passwords for different websites anymore. But this system not only has drawbacks in its efficiency and convenience as the third party-solution but also failed to tackle the trust issue. Li et al. [24–26] customized a genetic algorithm to solve the file block replica placement problem between multiple users and multiple data centers in the distributed cloud storage environment, but it only covers the perspective of genetic algorithm improvement on the network performance of the traditional architecture by reducing the costs of replica scheduling and transmitting. It lacks the coverage of security and performance.

Li et al. (2018) [24] proposed how it is possible to cope with Model-Driven Engineering techniques to security analysis and monitoring of cloud infrastructures. The methodology covers the entire life cycle of Infrastructure as a Service (Iaas) and provides a means to analyze systems at design and run-time; however, it does not cover the methodologies for complex governance of thermal and energy management. Shishido et al. [27] proposed Genetic-based Algorithms (GA) applied to a workflow scheduling algorithm with security and deadline constraints in cloud computing, solving the problem on optimization that examines the Particle Swarm Optimization (PSO) and GA in an attempt to optimize workflow scheduling. It provides approaches for the workflow scheduling process, but the optimization proposed is not simulated effectively.

Potey et al. [28] proposes that cloud computing consumers are permitted to depot a large volume of data onto the cloud for imminent usage. It directs the data in the enciphered form to be stored in the cloud with full homomorphism. The data is stored in DynamoDB of AWS public cloud. All the modifications/computations initiated by the user are executed on the enciphered data in a public cloud. After the computations are performed, the results of the computation — if required — can be downloaded on a local client machine [29–34]. To be precise, user data is not forced to be saved as plaintext on a public cloud. The main drawback of this paper is the size of the ciphertext space needed for efficient data management. An extra module is also needed to design search and query the enciphered data on the cloud when the homomorphic encryption technique is applied. Tosh et al. [8] proposed models and simulated the Block With Holding (BWH) attack in a blockchain-powered cloud, making an allowance for a unique pool reward scheme. It focuses on a particular model — the issue of block withholding attack — that is prevalent in PoW-based mining pools to understand the attacker's strategy toward taking over the pool members' rewards. Pay per Last N Shares (PPLNS) scheme could be useful in keeping the attacker's impact lesser than the proportional reward scheme.

In Pal et al. [35], a four-stage character/bit-level encoding technique has been proposed. Information is stored in databases and transmitted via complex networks. Hence, without appropriate protection, information is susceptible to unwanted exposures or unauthorized modifications, which may cause significant inconveniences. With the help of cryptography, messages are encoded with a multi-stage cipher technique, generating a ciphertext. It explains only the possibilities to enhance the method with assumptions and is not designed for an average case. Yu et al. [36–38] proposed an efficient privacy-preserving algorithm to preserve the privacy of information in social networks. It verifies their identities after they send a request. They make use of the recognition without tampering with the blockchain to store the user's public key and bind them to the block address, which is used for authentication. In this paper, they used only the primary hash function for better security but failed to cover high security and data remedial action, if data is leaked. Chen et al. [39] proposed the fifth-generation (5G) network to address the system capacity issue with Access Points Generation Practical Byzantine Fault Tolerance (APG-PBFT) algorithm based on blockchain technology used to improve the efficiency and reduce the authentication frequency of UE (User Equipment). This paper is based on Ultra-Dense Network (UDN), which is not adaptable for the existing 4G.

The fundamental concept behind cloud computing technology is virtualization that facilitates the maximum utilization of cloud resources. The primary concern of the virtual environment is the competency to host more virtual machines on limited physical machines so available computing resources can be shared. The principal challenge is the provision of a satisfactory level of service even at the time of outbursts in demand. Hence, there are crucial issues to be taken care of for a perfect realization in the present-day scenario. Of those, load balancing is the most prevalent subject, where there are virtual servers that are overloaded but, at the same time, leaving other resources to be under-utilized. There are several solutions put forth, like scheduling the requests and tasks, but it cannot resolve the difficulty of sharing the workloads in a distributed context. The task of assigning a job to the

Resource Optimization of Cloud Services 175

resource and balancing the workloads among the pool can be viewed as a combinatorial problem. To make a precise attempt to resolve, numerous branch and bound solutions and approximation techniques are proposed. However, these solutions cause an unsatisfactory outcome and impact in an extended time complexity that is not feasible for an outgrown cloud atmosphere (Table 10.1).

3 PROBLEM STATEMENT

Cloud storage architecture has a centralized management that gravely affects the security of the physical server and the demand for a trusted third-party authority to maintain the workloads. Even then, the prevalent system requires explicit algorithms to map the resources to the users optimally and to traceback the allocation. There is no point of validation in the access control policies on the data. The cloud platforms are consistently striving hard to improve the performance and costs tradeoff, but it poses a question on security concerns.

To provide more secure and reliable cloud storage services for enterprises or individual users with better performance on resource transactions is the primary objective. The concern is to impose the user to organize their data and configure finely-tuned access controls according to their business requirements. Entire transactional data can be queried and a transparent log of information obtained can help resource utilization and usage. Blockchain is used to protect personnel data through distributed storing file access permission in the cloud database system.

4 BI-LAYERING OF BLOCKCHAIN

The combinations of cryptography mechanisms and data integrity over the public ledger in a blockchain after a consensus is reached in the decentralized public

TABLE 10.1
Notations and Symbols Used

Notations and Symbols	Definitions
U	Set of cloud users
R	Set of cloud resources
PK{u}	Private key of cloud user
PU{u}	Public key of cloud user
PK{r}	Private key of cloud resource
PU{r}	Public key of cloud resource
K {u, r}	Attribute-based encryption key
APuv	Access policy defined on a user-resource mapping
T	Time instance
TC	Change in time, T
RMur	Resource request for measurement from the user
Pun	Public key of node n

ledger, allowing users to build any kind of application on top of the blockchain without worrying about trust components of the users and malicious behavior. The architectural design of the work is logically split into four layers — the resource layer and layers 0, 1, and 2 — as depicted in Figure 10.3.

Resource Layer: Limited space of physical machines hosted in the cloud service platform renders its resources to provide an infrastructure background for the execution of blockchain. It includes RAM, operating system, processor core, and system clock, and forms the backbone of the framework to build upon.

Private Peer to Peer Network Layer: Layer 0 is a P2P network where the blockchain nodes are connected to communicate data asynchronously. It comprehends the full-stack network of a conventional network architecture that concentrates on internet routing with reliable logical communication links between the nodes.

Main Blockchain Layer: Layer 1 hosts a permissioned blockchain that is a tamper-proof append-only database that holds several blocks of transactions that are linked for verification. These blocks are replicated among all the nodes to maintain the state of the chain.

a. Smart Contract: The piece of code written for imposing business logic that is invoked by an action. It triggers either a change of state or a microcomputation.
b. Consensus Model: It is a globally agreed-upon state of the participants by running a consensus algorithm to ensure the integrity. The classic types of consensus are Proof of Work (PoW) and Proof of Stake (PoS).

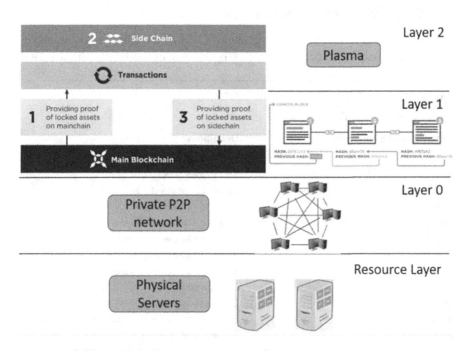

FIGURE 10.3 Bi-layered Blockchain Structure.

Sidechain Layer: A layer 2 logic allows transactions between the nodes by the exchange of authenticated data via a channel that is explicitly formed but tethered to the main blockchain. Authenticated transactions are submitted to the main chain only in case of any controversy, with a provision for the parent chain to finalize. The properties of the sidechain depend on the consensus algorithm of the main chain.

Software Environment: Node.js is an open-source server environment that uses JavaScript on the server. Truffle is used as a development environment, testing framework, and asset pipeline for Ethereum to get built-in smart contract compilation, linking, deployment, and binary management. Web3.js is a collection of libraries that allows us to interact with a local or remote Ethereum node using HTTP, IPC, or WebSocket. Ganache is used as a personal blockchain used to deploy contracts, develop the applications, and run tests. testrpc is a Node.js-based Ethereum client used for testing and development. It uses ethereum.js to simulate full client behavior as it includes all accessible RPC functions, events, and runs deterministically.

The pros of the bi-layered framework are that it minimizes the burden of the base layer and it hikes the throughput by an increased processing of transactions by data offloading.

5 SMART CONTRACTS

Smart contracts are programs or lines of code that get executed when the required conditions are satisfied; that is, it has some set of predefined rules. It removes the need for the participation of humans during the transaction process. Smart contracts use high-level languages such as **Solidity and** Viper. The smart contracts are customized by people and impose certain agreements on the transaction. It eliminates a need for a third person to verify and validate the transactions and is performed automatically. As they are executed automatically, it eliminates human error and provides better accuracy. The code is inserted in the blockchain. The transactions and contracts are recorded. Though the data in the code is evident on a public ledger, the privacy of the users remains secured. The smart contracts have properties similar to blockchain such as security, immutability, accuracy, and decentralization. Trading, supply chain, finance and data recording, and insurance are some use cases for smart contracts. Smart contracts have a particular agreement that drafts the rules and penalties; they reduce transaction costs and are conflict-free. Smart contracts eliminate the need for third-party validation, such as brokers or agents, and are convenient to use. It also provides a triggering event for an automated transaction. Smart contracts are referred to as self-executing, automated contracts. *Automation* and *cost-saving* are the most critical key factors in smart contracts. All transactions are automatically recorded, providing higher resistance to forgery.

6 SOLIDITY

To implement smart contracts that run on the Ethereum and other private blockchains, we use Solidity, which is an *object-oriented language*. Solidity has many use cases, but we mainly create contracts for voting. Solidity is performed to target

Ethereum Virtual Machine (EVM) and hold the concept of inheritance. The solidity program is run on *Remix IDE*. A Remix is a browser tool where contracts are written, deployed, and run. The concept of inheritance is also used in Solidity. Once the contract has been finished, no new features or updates can be made to it.

7   SYSTEM DESIGN

The design utilizes the conceptual framework of two-layered blockchain solutions to accommodate the optimization and planning of resources. At an abstraction level, the entities are grouped as cloud users, cloud service platforms, back-end blockchain, and tracking entity.

Procedure

1. User request	: The user initiates the request for remote access of the resources.
2. Define access control policy	: Data owners (users) can define the access privileges to their data.
3. Resource measurement data	: The data induced for resources to be transmitted in the cloud.
4. Data stock	: The physical nodes counter for directing the resource measurement data and access privileges onto blockchain.
5. Validation	: The administrators and authorized users are verified against the respective access privileges for allowance.
6. Data delegation	: Only valid admins and users are rendered with the measurement data.

Cloud users are the registered service requestors who are authorized for remote access to the offered services. Cloud service platforms assist in rendering storage, computing, and validation services to customers. These sources are executing on top of the physical machines. The users may be interested in fetching the usage histories of the targeted resource to plan; those who wish to access the tracing option can connect through the interface. Blockchain is utilized to save the placements of physical machines and the location of virtual servers on top of recording their usage information. Blockchain occupies the infrastructure of physical hosts to be maintained. The smart contract instantiated on top of the blockchain invokes the function that decides on the optimal resource allocation strategy based on the preceding resource-user mapping pattern. The flow is summarized in Figure 10.4.

Let the set of cloud users be $U = \{u1, u2, u3, ..., un\}$ and the resource set $R = \{r1, r2, r3, ..., rn\}$. For user authorization, public-key cryptography is utilized, and its private key $PK\{u1\}$ signs the resources allocated to the user $u1$. The public key $PU\{u1\}$ is used to locate the resource and validate it. To uniquely identify a resource, again, public-private key pair is used, and their corresponding private key $PK\{r1\}$ signs the measurements of all resources. The public key $PU\{r1\}$ is used to locate the resource and validate it. An attribute-based encryption key $K\{u1, r1\}$ is generated based on the access policy $APuv$ defined by the user $u1$ on the resource $r1$ on that particular instance of time T, and this policy can be redefined by the user U at any point of time TC (time change). The process of validation is listed out in the

Resource Optimization of Cloud Services

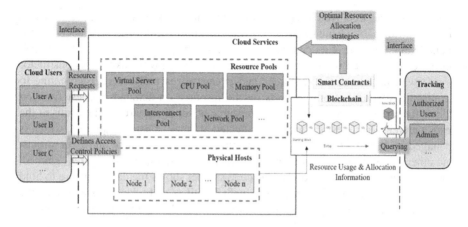

FIGURE 10.4 System Architecture.

algorithm below, where RMur is the resource r1 request for its measurement from the user u1. PUn denotes the public identity of the node n in the network.

Algorithm 1   User – Resource mapping validation

Procedure **Validate (RMur, APur(TC), PUr, PUu, PUn)**
   flag=0;
   RMur is a new request
   APur(T), PUr, PUu are public details
   if (Hur(t) = HASH(APur(TC) && T > TC))
     Mapping (RMur, APur) is valid
   else
     set flag=1
     return invalid
   return flag
End procedure

---

8   IMPLEMENTATION

Ethereum Plasma cuts down unnecessary computations on the main chain. It deals with the smart contracts and broadcast the verified transactions to the parent chain for further validation. This subprocess minimizes the processing power and time, thereby speeding up the transactions to get committed faster. Ganache sets up a local blockchain network that allows developers to create and deploy smart contracts and develop decentralized applications (Dapp). Truffle framework has been used to develop Dapps. The basic folder structure contains contracts, migrations, and truffle.js. Web3.js is used for interaction between smart contract and blockchain network. Metamask extension is used for browser support to connect with blockchain. The deployed smart contracts are stored as blocks in the blockchain network with a block number. Since they are now deployed, they become immutable and their business cannot be updated or changed further. The set of cloud resources is assigned with a unique identifier associated with tokens. A constructor is one who initializes the resources and decentralizes users to register online to acquire the tokens; set out requests on the resource on-demand; request function allows the user to place a request on a resource or set of resources. The status of the resource and the requestor should be reflected at any instance on queried, which is managed by the function Stats. The smart contract has Resource and User structs, and other data items — an array of items, an array of users, an array of req_granted, mappings, and physical locations. The code snippet shows the data structure to hold the details of cloud resources and users with identities and tokens. As a sample, the functions for a request to a resource and the users who maintained that resource.

```
struct Resource {
    uint resouceId;
    uint[] itemTokens; }
struct User {
    uint remainingTokens;
    uint userID;
    address location; }
function request (uint resorceID) public returns (uint) {
    require (resourceID > = 0 && resourceID < = threshold);
    users[resourceID] = msg.sender;
    return resouceID; }
function getUsers() public view returns (address[threshold] memory)
    { return users; }
```

Sample smart contract for optimal resource provisioning.

9 CONSENSUS ALGORITHM

The consensus mechanism aids in the attainment of agreement in the context of distributed computing environment [40]. There are two classes of classical models: Proof of Work (PoW) and Proof of Stake (PoS). The traditional PoW requires a high computing effort that would cause energy exploitation [41]. PoS is favorable only for the wealthy node and might cause starvation.

Conventional PoW: The poW algorithm is an original consensus mechanism used to verify the transactions and produce new blocks in the blockchain. It lets every node in the network solve a cryptographic puzzle to add a block. When a node finds the required nonce value, it is broadcasted to the entire network, and every node verifies it and stores the block in the blockchain as the next block. It is essential to the mining process that drives a cryptocurrency, where the distributed digital ledger at the core of a cryptocurrency is updated by nodes that are required to perform costly PoW computations. Mining is a complex process as the nodes need to demonstrate that they can validate a transaction block. Thus, complexity increases over time, and the power consumption also increases. As an illustration, let us consider a cryptographic puzzle where using the SHA-512 hash function, a node must find a target hash with the difficulty level $D = 5$,

$$SHA - 512 (block + Nonce) = 00000a45b668f34\ldots\ldots\ldots e3453$$
(Hash Value starting with "00000")

where the nonce value is an arbitrary number operated by the miner to solve the cryptographic hash puzzle that provides the miner with the right to broadcast the block.

Modified PoW: The proposed PoW-based consensus minimizes the energy consumption by restricting the number of nodes to take part in the consensus. All the physical nodes are sorted in the decreasing order of their current workloads. The miners of the previous β blocks are eliminated from the line. The nodes that stand in the first α positions of the line gets the chance to mine the next block. β is the mining control parameter, and α imposes the fairness in the consensus. The value of β should be optimally decided as per the code of the smart contract so that the heavily loaded physical host is not allowed to choose for mining at that instance.

Algorithm 2   Modified PoW Consensus

Based on α and β values, n nodes are selected
for all n nodes {
def generation(data):
    nonce = random.choice(ascii_lowercase+uppercase+digits)
    attempt = data + nonce
    return attempt,one
def proofOfWork(data):
    shash = hashlib.sha512()
    Found = False
    while Found == False:
        attempt, temp_nonce = generation(data)
        shash.update(attempt)
        solution = shash.hexdigest()
        if solution.startswith(difficultyLevel):
        nonce = temp_nonce
        Found = True
    data = "from:service_platform,To:user"
    proofOfWork(data)
}

9.1 Mathematical Analysis on Modified PoW

The parameter β has the control over mining. If the value of β is more significant, then there is a high probability for the heavy-loaded host to become a miner. Let us assume that each physical host has a probability τ to be malicious, the failure probability of mining is given by

$$P(Failure) = P(b(x; m, (1-\tau))) \leq \beta$$

where m is the number of physical hosts; $P(b)$ is the binomial probability that an m-trial binomial experiment that results in an exact x success in which the success probability of an individual trial is $(1-\tau)$. In the experimental environment, the number of nodes allowed for mining gets reduced by 64% ($\alpha = 6$) when $m = 20$ nodes. Hence, it shows that the energy exploitation of 19 nodes is to be cut down to 5 nodes. The given plot in Figure 10.5 depicts the comparison of conventional PoW with the modified PoW consensus in terms of the number of nodes given with the chance to take part in the mining process.

10 APPLICATION OF PROPOSED SOLUTION ON AWS S3

AWS S3 (Simple Storage Service) provides more flexible and reliable cloud storage services for enterprises or individual users. S3 is an object storage used to store and retrieve data from anywhere at any time. More real-time applications use AWS S3 like websites, mobile apps, IoT (Internet of Things) sensors, and so on. AWS Lambda offers services to run the code without a server, hence, the provision of a serverless computing environment. It can scale automatically to deal with more requests per second. It has been utilized to execute the back-end blockchain services. A lambda function is created to process the binary file payload to be directed

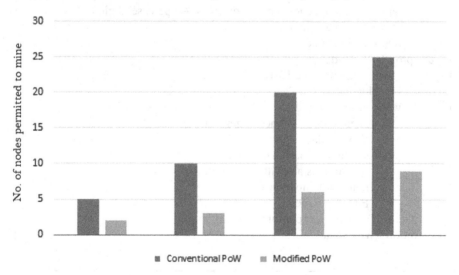

FIGURE 10.5 Comparison Plot between Conventional and Modified PoW.

Resource Optimization of Cloud Services

to the S3 bucket. Figure 10.6 exposes the sequential flow of the queries and responses from the entities, and Figure 10.7 shows the processed data view.

The Application Programming Interface (API) should be customized and adequately configured to acquire binary file types so that lambda can deserialize the payload. The store-to-be file are converted to binary files and are Base64 encoded to be parsed into lambda function via event payload structure. The binary streams are decoded to the original file to be saved to the S3 bucket. The gateway enables the uploading of files with the PUT method and retrieval of files with the GET method. Identity and Access Management (IAM) roles associated with the file are used for the PUT and GET methods to restrict access to the file upload bucket. Blockchain generates a key pair for each user to register their identity. The personal information is saved as hash values as a finger-print of sensitive data. In any instance, the user can request the administrator to verify the hash values by authenticating that the information stored on the blockchain is valid. Hence, whenever someone requires a user's identity for any kind of authentication or identification, they can use the hashes of the block pre-verified by the trusted, recognized administrator.

The users are permitted to define the access privileges for their data and information. The requisites are broadcasted to the blockchain nodes, which are responsible for validating and recording the usage measurements. The smart contract

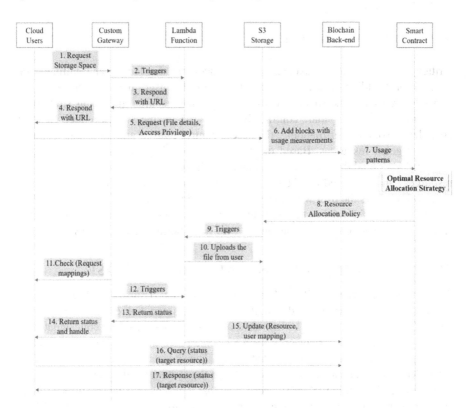

FIGURE 10.6 Proposed Solution on AWS S3 Storage.

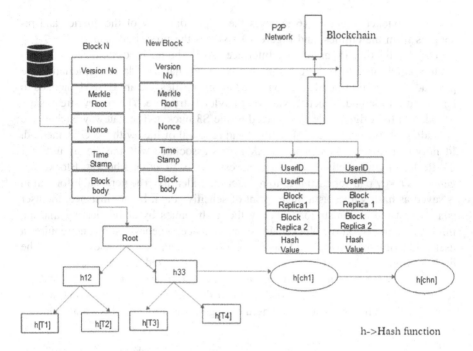

FIGURE 10.7 Processed Data View.

written on top of the blockchain network acquires the ledger and initiates the process of building a usage pattern for both the user and the resources independently as a background process. An optimal resource allocation strategy is recommended by the smart contract based on the business logic coded in it. It considers resource-user mapping, the access control matrix, and the usage histories of the targeted resources. Based on the suggested policy, the lambda function is triggered to upload the file into S3 storage. The history and the present status of any resources can be queried from the chain by an authorized user. AWS IAM is a web service that offers the user with a service to maintain secured access control to AWS resources. It differentiates the authenticated users who are signed in to use the resource from the authorized customers who have permission to access resources.

11 SIGNING AND AUTHENTICATING REST REQUEST

The Amazon S3 REST API needs a custom Hypertext Transfer Protocol (HTTP) scheme that is based on a keyed-HMAC (Hashed Message Authentication Code) for authentication purposes. To authenticate a request, users need to concatenate the selected elements of the request to form a string. Then, the AWS secret access key is used to arrive at the HMAC of that string. The digest is the signature, which is appended as a parameter of the request. When the system receives an authenticated request, it fetches the AWS secret access key that it claims to have and uses it, in the same way, to compute a signature for the received message. It then compares the signature it arrived at against the signature that is shown by the requester. If the two signatures match, then the system concludes that the requester must have access to the AWS secret access key and, therefore, acts with the authority of the principal to whom the key was issued. If the two signatures do not match, the request is dropped, and the system responds with an error message.

Public access is granted to buckets and objects through Access Control Lists (ACLs), bucket policies, or both. To help the user manage public access to Amazon S3 resources, Amazon S3 provides block public access settings. Amazon S3 block public access settings can override ACLs and bucket policies so that users can enforce uniform limits on public access to these resources. Users can apply block public access settings to individual buckets or to all buckets in the AWS account. To help ensure that all Amazon S3 buckets and objects have their public access blocked, all four settings for block public access for the AWS account are recommended to be turned on. These settings block public access from all current and future buckets.

Authorization with AWS access key for signature

1. Authorization = "AWS" + " " + AWSAccessKeyId + ":" + Signature;
2. Signature = Base64(HMAC-SHA1(SecretAccessKey,UTF-8-Encoding-Of (StringToSign)));
3. StringToSign = HTTP-Verb + "\n" +
    i. Content-MD5 + "\n" +
    ii. Content-Type + "\n" +
    iii. Date + "\n" +
    iv. CanonicalizedAmzHeaders +
    v. CanonicalizedResource;
4. CanonicalizedResource = ["/" + Bucket] +
    i. <HTTP-Request-URI, from the protocol name up to the query string> +
    ii. [subresource, if present. For example "?acl", "?location", "?logging", or "?torrent"];
5. CanonicalizedAmzHeaders = <labelled in the table>

Access Key and Security Key

Parameter	Value
AWSAccessKeyId	JKSHDSDDSLKFJLFEXAMPLE
AWSSecretAccessKey	ejfefFSDFFSDF/KKSDFJ32KFCMSD/bPxRfiSDFHSDK

12 RESULTS

The outcome of the work is discussed concerning some critical performance metrics. Figure 10.8 depicts the correlation between the number of requests and the processing

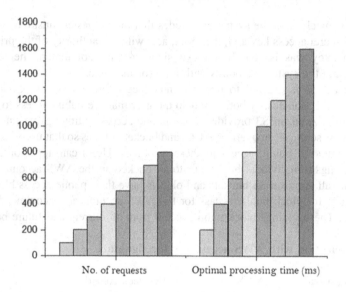

FIGURE 10.8 Optimal Processing Time of Blocks.

time of the blocks. As a two-layered structure has been designed, the processing time is found to be just twice as that of the requests raised per millisecond.

Figure 10.9 shows the comparison of the proposed solution implemented in AWS S3 storage with the competing platform Microsoft Azure. The graph is plotted for the latency against the time. The latency has been viewed to be very minimal.

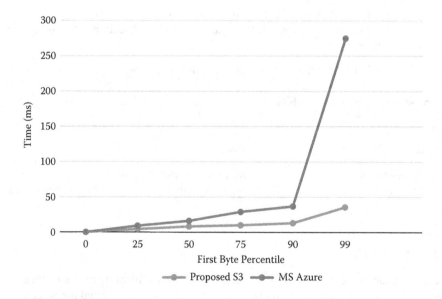

FIGURE 10.9 Time to First-Byte Latency in the Proposed Context.

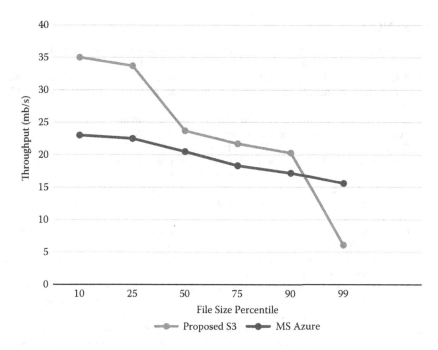

FIGURE 10.10 Throughput of Bi-layered Blockchain.

Hence, the optimization achieved has an impact on the latency of the requests and responses.

Figure 10.10 shows that the throughput hikes even though the system is dumped with extensive voluminous data and does not seem to reach the bottleneck as the file size to be uploaded increases. It is found that the proposed solution has a high impact on the throughput for the amount of data to be stored.

13 CONCLUSION

With the advancements in technologies, systems have become sophisticated over the last couple of years. Therefore, it is extremely crucial to secure data by some mechanisms to preserve data integrity. The blockchain network has risen as an alternate option to focus on performance upgrades and a solution to security risks in a distributed system. In this work, the decentralized security architecture based on a two-layered blockchain has been proposed and compared with other distributed cloud storage. This system has been stated with better security and performance with essential infrastructure. The work has been proven to be useful in the S3 storage.

REFERENCES

1. Rajathi and Saravanan (2013), A survey on secure storage in cloud computing, *Indian Journal of Science and Technology*, 6(4), pp. 4396–4401.
2. Bohme, R., Christin, N., and Edelman, B. (2015) Bitcoin: Economics, technology and governance, economic perspectives, *Journal of Economic Perspectives*, 29(2), pp. 213–38.
3. Zhang, R., Lin, C., Meng, K., and Zhu, L. (2013), A modeling reliability analysis technique for cloud storage system, in IEEE International Conference on Communication Technology (ICCT), IEEE (pp. 32–36).
4. Nakamoto, S., and Bitcoin, N. (2008), A peer-to-peer electronic cash system, *Indian Journal of Science and Technology*, 9(4), pp. 4896–4601.
5. Chen, S., Irving, S., and Peng, L. (2016) Operational cost optimization for cloud computing data centers using renewable energy, *IEEE Systems Journal*, 10(4), pp. 1447–1458.
6. Johnson, D., Menezes, A., and Vanstone, S. (2001) The elliptic curve digital signature algorithm (ECDSA), in *International Journal of Information Security*, 1(1), pp. 36–63.
7. Ethereum Foundation. (2017, September 18). Ethereum, white paper, https://github.com/ethereum/wiki/wiki/White-Paper.
8. Tosh, D., and Shetty, S. (2018) Security implications of block chain cloud with analysis of block withholding attack, in Proceedings of the 17th IEEE/ACM International Symposium on Cluster, Cloud, and Grid Computing, IEEE Press pp. (458–467).
9. Kecong, D., Tse, H., Cai, B., and Liang, K. (2018). Robust password-keeping system using block-chain technology, in Proceedings of the 17th IEEE/ACM International Symposium on Cluster, Cloud, and Grid Computing, IEEE Press (pp. 458–467).
10. Casino, F., Dasaklis, T. K., and Patsakis, C. (2019) A systematic literature review of blockchain-based applications: Current status, classification and open issues, *Telematics and Informatics*, 36, pp. 55–8180 DOI 10.1016/j.tele.2018.11.006.
11. Buldas, A. K., and Laanoja, R. (2013). Keyless signatures infrastructure: How to build global distributed hash-trees, in Nordic Conference on Secure IT Systems (pp. 313–320).
12. Chandra, P. J., Ponsy, R. K., Bhama, S., Swarnalaxmi, S., Safa, A. A., and Elakkiya, I. (2018), Blockchain centered homomorphic encryption: A secure solution for E-balloting, in *Lecture Notes on Data Engineering and Communications Technologies*, Springer.
13. Priya, J. C. and RK, S. B. P. (2018, December). Disseminated and decentred blockchain secured balloting: Apropos to India, in *Proceedings of IEEE International Conference on Advanced Computing (ICoAC)* (pp. 323–327). IEEE.
14. Ariel, E., Azaria, A., Halamka, J. D., and Lippman, A. (2016) A case study for blockchain in healthcare:medrec prototype for electronic health records and medical research data, in Proceedings of IEEE Open & Big Data Conference (vol. 13, pp. 13).
15. Christidis, K., and Devetsikiotis, M. (2016) Blockchains and smart contracts for the internet of things, *IEEE Access*, 4, pp. 2292–2303.
16. Heilman, E., Baldimtsi, F., and Goldberg, S. (2016) Blindly signed contracts: Anonymous on-blockchain and off-blockchain bitcoin transactions, in International Conference on Financial Cryptography and Data Security (pp 3–60). Springer.
17. Bogumił, K., and Szufel, P. (2015) On optimization on simulation execution on Amazon EC2 spots Market, *Journal of Simulation Modelling Practice and Theory*, 58, pp. 172–187.
18. Al Omar, A., Bhuiyan, M. Z. A., Basu, A. et al. (2018) Privacy-friendly platform for healthcare data in cloud based on blockchain environment, *International Journal of Information Security*, 95, June 2019, pp. 511–521.

19. Dagher, G., Mohler, J., Milojkovic, M., and Ancile, P. M. (2018) Privacy-preserving framework for access control and interoperability of electronic health records using blockchain technology, *Journal of Sustainable Cities and Society*, *39*, pp. 283–297.
20. Zhang, Z., Qin, Z., Zhu, L., Weng, J., and Ren, K. (2017). Cost-friendly differential privacy for smart meters, *IEEE Transactions on Smart Grid*, *8*(2), pp. 619–626.
21. Wang, X., Yang, W., Noor, S., Chen, C., Guo, M., and van Dam, K. H. (2019). Blockchain-based smart contract for energy demand management, *Energy Procedia*, *158*, pp. 2719–2724.
22. Zyskind, G., Nathan, O. et al., (2015). Decentralizing privacy: Using blockchain to protect personal data, *IEEE Security and Privacy Workshops (SPW)*, pp. 180–184.
23. Gungor, V. C., Sahin, D., Kocak, T., Ergut, S., Buccella, C., Cecati, C., and Hancke, G. P. (2013) A survey on smart grid potential applications and communication requirements, *IEEE Transactions on Industrial Informatics*, *9*(1), pp. 28–42.
24. Li, J., Liu, Z., and Guangdong (2018), Block chain-based security architecture for distributed of IaaS resources, in Proceedings of IEEE/ACM International Conference Grid Computing (GRID) (pp. 9–16).
25. Choe, J., and Yoo, S. K. (2008) Web-based secure access from multiple patient repositories, *International Journal of Medical Informatics*, *77*(4), pp. 242–248.
26. Gai, K., Zhu, L., Wu, Y., and Choo, K.-K. R. (2018) Controllable and trustworthy blockchain-based cloud data management, *Journal of Future Generation Computer Systems*, *91*, pp. 527–535.
27. Shishido, H. Y. (2017), Genetic-based algorithms applied to a workflow scheduling algorithm with security and deadline constraints in cloud computing, *Indian Journal of Science and Technology*, *6*(4), pp. 4396–4401.
28. Potey, M. M., and Dhote, C. A. (2017) Homomorphism encryption for security of cloud data, *Journal of Procedia Computer Science*, *79*, pp. 175–181.
29. Tebaa, M., El Hajji, S. and El Ghazi, A. (2012), Homomorphic encryption method applied to cloud computing, in *International Journal of Information & Computation Technology*, *4*(15), pp.1519–1530.
30. Le-Dang, Q. and Le-Ngoc, T. (2019), Scalable blockchain-based architecture for massive IoT reconfiguration, in 2019 IEEE Canadian Conference of Electrical and Computer Engineering (CCECE). IEEE (pp. 1–4).
31. Lee, C., Kim, H., Maharajan, S., Ko, K., and Hong, J. W.-K. (2019), Blockchain explorer based on RPC-based monitoring system, in 2019 IEEE International Conference on Blockchain and Cryptocurrency (ICBC). IEEE (pp. 117–119).
32. Lee, H., Sung, K., Lee, K., Lee, J., and Min, S. (2018). Economic analysis of blockchain technology on digital platform market, in 2018 IEEE 23rd Pacific Rim International Symposium on Dependable Computing (PRDC). IEEE (pp. 94–103).
33. Banno, R., and Shudo, K. (2019). Simulating a blockchain network with simblock, in Proceedings of the 2019 IEEE International Conference on Blockchain and Cryptocurrency (IEEE ICBC 2019) (pp. 3–4).
34. Zhang, Y., Wang, S. and Zhang, Y. (2017). A block chain-based framework for data sharing with fine-grained access control in decentralized storage systems, *Proceedings of IEEE Access*, *6*, pp. 38437–38450.
35. Pal, J. K., and Mandal, J. K. (2009), A random block length based cryptosystem through multiple cascaded permutation combinations and chaining of blocks, in International Conference on Industrial and Information Systems (ICIIS) (pp. 26–31).
36. Yu, M. (2018), Authentication with block-chain algorithm and text encryption protocol in calculation of social network, *IEEE Access*, *5*, pp. 24944–24951.
37. Koshy, P., Koshy, D., McDaniel, P. (2014), An analysis of anonymity in bitcoin using p2p network traffic, in International Conference on Financial Cryptography and Data Security (pp. 469–485).

38. Ramaiah, Y. G., and Kumari, G. V. (2012) Efficient public key homomorphic encryption over integer plaintexts, in International Conference on Information Security and Intelligence Control (ISIC), IEEE (pp. 123–128).
39. Chen, Z., Chen, S., Xu, H., and Hu, B. (2017). A security authentication scheme of 5G ultra-dense network based on block chain, *IEEE Access*, 6, pp. 55372–55379.
40. Zhang, S., and Lee, J. H. (2019). Analysis of the main consensus protocols of blockchain, *ICT Express*, DOI 10.1016/j.icte.2019.08.001.
41. Cao, B., Zhang, Z., Feng, D., Zhang, S., Zhang, L., Peng, M., and Li, Y. (2019). Performance analysis and comparison of PoW, PoS and DAG based blockchains, *Digital Communications and Networks*, DOI 10.1016/j.dcan.2019.12.001.

11
Trust-Based GPS Faking Attack Detection in Cognitive Software-Defined Vehicular Network (CSDVN)

Rajendra Prasad Nayak[1], Srinivas Sethi[2], and Sourav Kumar Bhoi[3]

[1]Department of Computer Science and Engineering, Government College of Engineering, Kalahandi, India, Email: rajendra.cet07@gmail.com

[2]Department of Computer Science Engineering and Applications, Indira Gandhi Institute of Technology, Sarang, India, Email: srinivas_sethi@igitsarang.ac.in

[3]Department of Computer Science and Engineering, Parala Maharaja Engineering College, Berhampur, India, Email: sourav.cse@pmec.ac.in

1 INTRODUCTION

Vehicular communication technology is largely rising to provide safety and non-safety applications to end-users [1–4]. As accidents grow every day, it is a problem for the government and private organizations to save the life of humans. Therefore, vehicular ad hoc network (VANET) provides a platform of manual, semi-automation, and full automation technology. By using the technologies, the vehicles in the network communicate valuable information with neighbors, other vehicles, RSUs, BS, local controller, and SDN controller. For this, communication vehicles use vehicle to vehicle (V2V) and vehicle to infrastructure communication (V2I). Vehicles in the network are highly dynamic; therefore, the vehicles use position-based routing to transfer the information to other nodes for reliable communication.

Nowadays, Software Defined Network (SDN) provides a better architecture for VANET, which improves the performance of the network [5–8]. SDN is a centralized architecture that divides the network into a data plane and control plane to provide services like QoS, security, optimization, low cost, availability, and better resource utilization in a dynamic environment. It makes the network flexible where the total control of the network is with the SDN controller at the control end. From the control end, the SDN controller controls the whole network by setting rules and regulations. The two layers are connected by the Openflow protocol. The two technologies are hybridized to generate a Software-Defined Vehicular Network (SDVN). The SDVN network divides the network into the data plane and control plane. In the control plane, the SDN controller exists and in the data plane, the vehicle—that has small intelligence for processing data—exists, and it follows the policies set by SDN to execute all works. This will be a better technology to divide the work into both parts. However, if the vehicles use cognitive computing technology, then it will be better for the network to perform well. Cognitive computing [9–11] facilitates the SDVN to work better by using techniques such as AI, machine learning, reasoning, etc. The techniques help predict, classify, etc. by taking prior knowledge about the network behavior. This concept motivates us to build a CSDVN network.

Security is a major concern in urban VANETs, as many types of attacks to control the network and reduce network performance (by disturbing the network applications) are possible nowadays [8,12–40]. GPS faking attack [12,13] is a type of attack where a vehicle in the network fakes the original GPS position by providing an incorrect GPS position in the beacon signals. This attack needs to be detected in the network for proper functioning.

The major contributions in this work are stated as follows:

1. A trust-based GPS faking attack detection method is proposed for cognitive software defined vehicular network (CSDVN). In this method, the vehicles in the network use the received signal strength indicator (RSSI) value of the beacon signals of neighbor vehicles to detect malicious activity.
2. The vehicle that directly tests its neighbor's malicious activity tags them with a direct trust value, and if the test is done by its neighbor vehicles upon another neighbor's vehicle, then it uses those values to tag that neighbor vehicle with an indirect trust value. These two trust values are combined to determine the exact trust of a neighbor's vehicle.
3. The trust values of the neighbor vehicles are sent to the RSU, then the RSU sends the trust values to the local controller (LC) under SDN through the base station (BS) to generate the final current trust value. The LC applies the cognitive concept to use the prior trust with received trust to generate the final trust. If the final current trust value is less than a threshold, then the vehicle is considered malicious; otherwise, it is genuine.
4. The performance of the method is evaluated using OMNET++ by considering various network parameters.

The rest of the work is distributed as follows. Section 2 presents related works. Section 3 presents the trust-based GPS faking detection method in CSDVN. Section 4 presents the simulation and results. Section 5 presents the conclusion and future scope.

Trust-Based GPS Faking Attack Detection 193

2 RELATED WORKS

Many works have been done to detect the GPS faking attack in the network. However, we have used the CSDVN network that gives a novel approach to the whole work. The attack in this network is also checked using our detection mechanism. Researchers mainly work in SDVN, VANET, MANET, WSN, IoT, etc. to detect position cheating attacks. The GPS faking attack is a very common attack in vehicular networks and many works have been done by researchers to detect this. Many such related works are presented in [8,12–40].

3 TRUST-BASED GPS FAKING ATTACK DETECTION METHODOLOGY FOR CSDVN

This section describes the network model, the misbehavior model, and the proposed approach. The network model describes the network devices and communication between them, shown in Figure 11.1. The misbehavior model describes the misbehavior performed by the attacker vehicles in the network. The proposed method discusses the GPS faking attack detection method in CSDVN.

3.1 NETWORK MODEL

Vehicle: The network consists of vehicles (V) with an onboard unit (OBU) in the city area. The vehicle can send, receive, and route packets. It has processing, storage, intelligence capability, and high battery status. The vehicle can communicate with other vehicles or RSU using V2V and V2I communication through wireless links in an ad hoc fashion. A vehicle beacons its current position, destination, direction, speed, and trust value in a particular period in its communication range. The current position is calculated using GPS and maps at the vehicle OBU unit. The beacon structure is shown in Figure 11.2 The vehicle has a neighbor table where it updates the data it receives from another vehicle. A vehicle can be attacked by another vehicle or a group of vehicles in its neighbor. Vehicles use a position-based routing protocol to send the data to a destination. The vehicle farthest in the communication range is selected as the next vehicle. The vehicle uses the Euclidean distance technique to find the next vehicle. The distance is calculated using the following formula as follows:

$$d = \sqrt{(x1 - x2)^2 + (y1 - y2)^2} \quad (11.1)$$

where d is the distance between the two vehicles, and (x1,x2) and (y1,y2) are the coordinates of the two vehicles in communication range.

1. **RSU:** RSU can send, receive, and route the packets to the destination. RSU is set at the junction to record the information (beacon) of the vehicles moving from one junction to another. It has processing, storage, intelligence capability, and high power status. RSU communicates with the vehicles using I2V communication and with the BS using Wimax/LTE connection. RSU

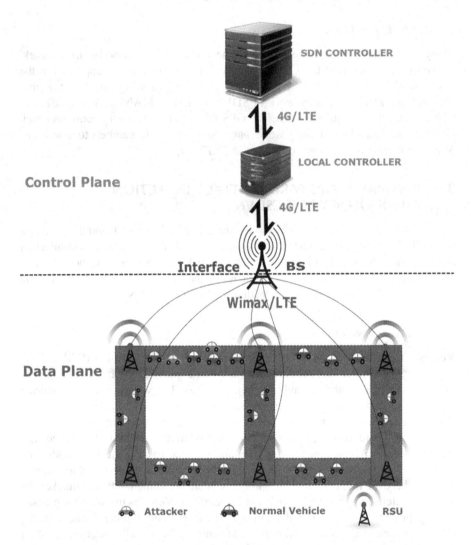

FIGURE 11.1 Proposed Framework to Detect the GPS Faking Attack Detection in CSDVN.

is a genuine node. It forwards the trust values of the suspected vehicles (untrusted vehicles information received from other vehicles) to the LC through BS. It also advertises information when required, for safety and non-safety purposes.

2. **BS:** The BS is responsible for receiving and transferring data from one layer to another (data plane to control plane and vice versa). It has processing, storage, intelligence capability, and high power status.

3. **Interface:** Interface is the layer that connects the two planes by a BS. The BS acts as an interface that transfers data packets and control packets between

| source Vid | position | destination | direction | speed | trust value |

FIGURE 11.2 Packet Format for Beacon.

layers. It also sends flow rules and policy rules from the control layer to the data layer.
4. **LC:** The LC at the control plane is responsible for processing, receiving, and transferring data to the SDN controller or BS. It has processing, storage, and intelligence capability. It has high power status. LC helps SDN by performing smaller tasks and sending bigger tasks to the SDN controller. Here, we consider the final trust computation as a smaller task performed at LC. LC also stores the previous trust values that are updated.
5. **SDN Controller:** The SDN controller at the control plane runs software that manages all the work in the CSDVN network. It defines the above policies and rules, and flow rules for the entire network devices in the network. It has high processing, high storage, high intelligence capability, and high power status. SDN stores the GPS faking attackers information that are constantly updated.

3.2 Misbehavior Model

In this work, we have mainly focused on a misbehavior performed by the vehicles in the CSDVN network. The vehicles in this network performs GPS faking attack. In this attack, the attacker vehicle fakes the GPS position in the beacon to get access over the network and perform network disruption activities. Figure 11.3 shows the GPS faking attack, where vehicle V sends the incorrect position in front of other neighbor vehicles; the vehicle V is behind other vehicles. This is a serious attack and needs to be detected early.

3.3 Proposed Methodology for GPS Faking Attack Detection

In this section, we described the method and algorithm of the whole network. All vehicles beacon a short message to its neighbors and a neighbor table is maintained at all vehicles. Let—after receiving the beacon message—a vehicle V1 check the

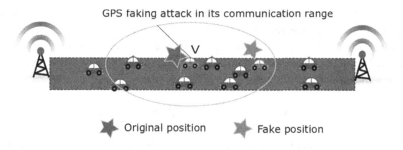

FIGURE 11.3 GPS Faking Attack by the Attacker Vehicle V in CSDVN.

correct position of its neighbor vehicle V2 and set a local trust value for that vehicle. This is called direct trust DT. It also receives the local trust values calculated by its other neighbors for vehicle V2. The values are given by the common neighbors of V1 and V2. V1 uses the values to calculate the indirect trust IT. The direct and indirect trust are combined to calculate the current LTV of V2. The values are calculated using the following equations as follows:

$$DT = \begin{cases} 1, & if\ dgps = =drssi \\ 0, & else \end{cases} \quad (11.2)$$

Where dgps is the distance calculated by using Equation (11.1) by taking the received position. drssi is the distance calculated by taking the beacon signal RSSI value as follows:

$$drssi = 10^{\left(\frac{measured\ power - rssi}{10 \times N}\right)} \quad (11.3)$$

where measured power is the power calculated by the vehicle at its end when the beacon is received, rssi is the power at 1 m, and N is a constant it ranges from 2 to 4.

Hence, from Equation (11.2) if dgps is the same as drssi, then the trust for that vehicle is 1; otherwise, 0 if malicious. The malicious vehicle sends an incorrect position; hence, the two distances are different. If vehicle V2 is suspicious, then vehicle V1 sends a message to its neighbor to send the DT values. After receiving the DT values from common neighbors, V1 finds IT as follows:

$$IT = max\,(frequency - 0,\ frequency - 1) \quad (11.4)$$

where IT is the value with maximum frequency of 0 and 1. Let, the IT of V2 be 0 because all its neighbors get DT=0 and it has high frequency. Hence, trust T for vehicle V2 is calculated as the combination of DT and IT as follows:

$$T = \alpha \times DT + (1 - \alpha) \times IT \quad (11.5)$$

where α is a weight parameter with value 0.5. The vehicles, after receiving the GPS, check the DT and if suspicion occurs, calculates the T for that vehicle. Then vehicle, V1 sends all suspected vehicles trust values T to the nearest RSU using position-based routing. After receiving the suspected vehicle information, RSU forwards the information to LC through BS. The LC contains the previous trust values of these suspected vehicles. These prior data are combined with the new trust value to get the final trust value as follows:

$$T = \beta \times Tcurrent + (1 - \beta) \times Tprevious \quad (11.6)$$

where β is a weight parameter with value 0.5. If the trust value is less than a threshold th, then the suspected vehicle is considered malicious and the information is sent to the RSU

> **Algorithm 1 Pseudocode for detection of GPS faking attackers in CSDVN.**
>
> **Input:** T values of suspected vehicles
> **Output:** Malicious vehicle
>
> 1. RSU sends T values of suspected vehicles to LC
> 2. LC computes $T = \beta \times Tcurrent + (1 - \beta) \times Tprevious$
> 3. **If** (T < th)
> 4. Malicious;
> 5. **Else**
> 6. Genuine;
> 7. **EndIf**
> 8. LC sends malicious information to SDN and RSU
> 9. RSU broadcasts the malicious information
> 10. **End**

and SDN controller by the LC. The RSU broadcasts the malicious vehicle's information in the network for proper functioning. The vehicles receive the data and inform their neighbors. Algorithm 1 shows the pseudocode for the proposed methodology.

4 SIMULATION AND RESULTS

In this section, the simulation and results are presented. Initially, the simulation setup and assumptions taken during the simulation are discussed. The parameters

TABLE 11.1
Simulation Setting for SUMO

Sl. No.	Parameters	Values
1	Area	3000×2000 m^2
2	Number of Lanes	3 × direction
3	Maximum Speed of Vehicles	30 m/s
4	Allowed Maximum Speed on Edges	30.556 m/s
5	Maximum Acceleration	3.0 m/s^2
6	Maximum Deceleration	6.0 m/s^2
7	Driver Imperfection	0.5
8	Number of vehicles	100
9	Vehicles type	3
10	Length of vehicle type	5 m, 7 m, 12 m

TABLE 11.2
Simulation Setting for Network Simulator

Sl. No.	Parameters	Values
1	Simulation time	300 s
2	Bitrate	6 Mbps
3	Packet generation rate	10 packets/s
4	Communication range of vehicle	300 m
5	Communication range of RSU	500 m
6	Update interval	0.1 s
7	IEEE	802.11p
8	Sensitivity	−80 dBm
9	Th	60%
10	SDN controller	1
11	Local controller	1
12	Number of RSU	4
13	BS	1
14	Number of simulations	20

considered for performance evaluation are detection accuracy, detection time, and energy consumption.

To carry out the simulation, we have used the Veins hybrid framework simulator. This simulator uses the IEEE 802.11p standard for communication. Since the framework is a hybrid, it uses OMNeT++ and Simulation of Urban Mobility (SUMO) as network and road traffic simulator, respectively [41]. These simulators are integrated using a Traffic Control Interface (TraCI) that provides the TCP connection between the network and road traffic simulator and maintains real-time

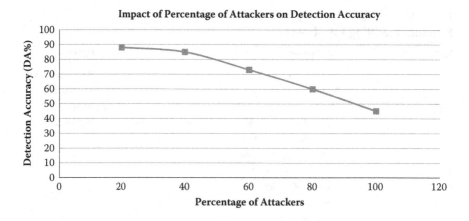

FIGURE 11.4 Detection Accuracy.

Trust-Based GPS Faking Attack Detection

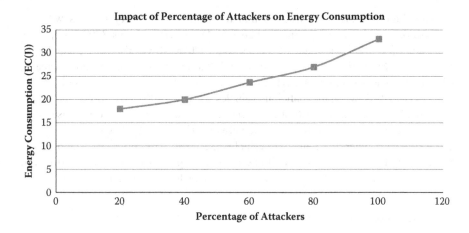

FIGURE 11.5 Energy Consumption.

interaction between them. We simulated our work in a grid map scenario. The configuration parameters for SUMO and OMNeT++ are provided in Table 11.1 and Table 11.2, respectively. The performance is evaluated by varying the malicious activity in the network. The percentage of malicious nodes is varied from 20-100% (16, 32, 48, 64, 80) of 80% malicious nodes of 100 vehicles. For example, 16 nodes are randomly set and tagged as malicious. The malicious nodes with coordinates (x,y) are changed with ±20 m error. The other 84 good nodes are also randomly deployed in the network. The initial trust value of malicious vehicles is set to 0.5 and for genuine node, to 0.8. The results are the average of 20 simulation runs.

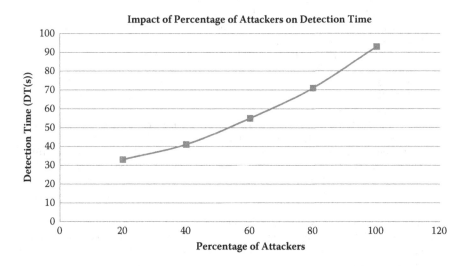

FIGURE 11.6 Detection Time.

From Figure 11.4, it is observed that the detection accuracy reduces with the increase of malicious vehicles in the network. However, it shows 88%, 85%, 73%, 63%, and 45% when the malicious vehicles are 20%, 40%, 60%, 80%, and 100%, respectively. From Figure 11.5, it is observed that the energy consumption increases with the increase of malicious vehicles in the network. However, it shows 18, 20, 23.7, 27, and 33 J, when the malicious vehicles are 20%, 40%, 60%, 80%, and 100%, respectively. From Figure 11.6, it is observed that the detection time increases with the increase of malicious vehicles in the network. However, it shows 33, 41, 55, 71, and 93 seconds when the malicious vehicles are 20%, 40%, 60%, 80%, and 100%, respectively. From this discussion, it is seen that the proposed method performs well by detecting the GPS faking attack in all parameter level conditions.

5 CONCLUSION

From the work above, it is found that the proposed GPS faking attack detection method can detect attacker vehicles with better accuracy and less detection time. The energy consumption of the nodes is found to be more for malicious vehicles. The proposed method will be a better algorithm to detect GPS faking attacks in WSN, VANET, SDVN, IoT, etc. networks. Simulation works show that this method can be implemented practically in a city area. In the future, we will implement this work using application-based software and hardware for better validation. The comparison will also be performed with other existing schemes.

REFERENCES

1. Hasrouny, H., Samhat, A. E., Bassil, C., and Laouiti, A. (2017). VANet security challenges and solutions: A survey. *Vehicular Communications*, 7, pp. 7–20.
2. Li, F. and Wang, Y. (2007). Routing in vehicular ad hoc networks: A survey. *IEEE Vehicular technology magazine*, 2(2), pp.12–22.
3. Cunha, F., Boukerche, A., Villas, L., Viana, A., and Loureiro, A. A. (2014). Data communication in VANETs: a survey, challenges and applications. *Network, IEEE Communications Surveys and Tutorials*.
4. Bhoi, S. K. and Khilar, P. M. (2013). Vehicular communication: A survey. *IET Networks*, 3(3), pp. 204–217.
5. Kreutz, D., Ramos, F. M. V., Verissimo, P. E., Rothenberg, C. E., Azodolmolky, S., and Uhlig, S. (2014). Software-defined networking: A comprehensive survey. Proceedings of the IEEE, *103*(1), (pp. 14–76).
6. Ku, I., Lu, Y., Gerla, M., Gomes, R. L., Ongaro, F., and Cerqueira, E. (2014). Towards software-defined VANET: Architecture and services. In 2014 13th Annual Mediterranean Ad Hoc Networking Workshop (MED-HOC-NET) (pp. 103–110). IEEE.
7. Truong, N. B., Lee, G. M., and Ghamri-Doudane, Y. (2015). Software defined networking-based vehicular adhoc network with fog computing. In 2015 IFIP/IEEE International Symposium on Integrated Network Management (IM) (pp. 1202–1207). IEEE.
8. Arif, M., Wang, G., Wang, T., and Peng, T. (2018). SDN-based secure VANETs communication with fog computing. In International Conference on Security, Privacy and Anonymity in Computation, Communication and Storage (pp. 46–59). Springer, Cham.

9. Marfia, G., Roccetti, M., Amoroso, A., Gerla, M., Pau, G., and Lim, J.-H. (2011, October). Cognitive cars: Constructing a cognitive playground for VANET research testbeds. In Proceedings of the 4th International Conference on Cognitive Radio and Advanced Spectrum Management (pp. 1–5).
10. Abuelela, M., and Olariu, S. (2010). Taking VANET to the clouds. In Proceedings of the 8th International Conference on Advances in Mobile Computing and Multimedia (pp. 6–13).
11. Modha, D. S., Ananthanarayanan, R., Esser, S. K., Ndirango, A., Sherbondy, A. J., and Singh, R. (2011). Cognitive computing. *Communications of the ACM, 54*(8), pp. 62–71.
12. Sakiz, F. and Sen, S. (2017). A survey of attacks and detection mechanisms on intelligent transportation systems: VANETs and IoV. *Ad Hoc Networks, 61*, pp. 33–50.
13. Kerrache, C. A., Calafate, C. T., Cano, J.-C., Lagraa, N., and Manzoni, P. (2016). Trust management for vehicular networks: An adversary-oriented overview. *IEEE Access, 4*, pp. 9293–9307.
14. Hu, H., Lu, R., Zhang, Z., and Shao, J. (2016) REPLACE: A reliable trust-based platoon service recommendation scheme in VANET. *IEEE Transactions on Vehicular Technology, 66*(2), pp. 1786–1797.
15. Lyamin, N., Vinel, A., Jonsson, M., and Loo, J. (2013). Real-time detection of denial-of-service attacks in IEEE 802.11 p vehicular networks. *IEEE Communications Letters, 18*(1), pp. 110–113.
16. Hu, H., Lu, R., Huang, C., and Zhang, Z. (2016). Tripsense: A trust-based vehicular platoon crowdsensing scheme with privacy preservation in vanets. *Sensors, 16*(6), pp. 803.
17. Amoozadeh, M., Deng, H., Chuah, C.-N., Zhang, H. M., and Ghosal, D. (2015). Platoon management with cooperative adaptive cruise control enabled by VANET. *Vehicular Communications, 2*(2), pp. 110–123.
18. Boeira, F., Barcellos, M. P., de Freitas, E. P., Vinel, A., and Asplund, M. (2017). Effects of colluding Sybil nodes in message falsification attacks for vehicular platooning. In 2017 IEEE Vehicular Networking Conference (VNC) (pp. 53–60). IEEE.
19. Asplund, M. (2014). Poster: Securing vehicular platoon membership. In 2014 IEEE Vehicular Networking Conference (VNC) (pp. 119–120). IEEE.
20. Patounas, G., Zhang, Y., and Gjessing, S. (2015). Evaluating defence schemes against jamming in vehicle platoon networks. In 2015 IEEE 18th International Conference on Intelligent Transportation Systems (pp. 2153–2158). IEEE.
21. Vitelli, D. (2016). Security vulnerabilities of vehicular platoon network. University of Naples Federico II Studies, Thesis.
22. Al Mutaz, M., Malott, L., and Chellappan, S. (2013). Leveraging platoon dispersion for sybil detection in vehicular networks. In 2013 Eleventh Annual Conference on Privacy, Security and Trust (pp. 340–347). IEEE.
23. Lyamin, N., Kleyko, D., Delooz, Q., and Vinel, A. (2018). AI-based malicious network traffic detection in VANETs. *IEEE Network, 32*(6), pp. 15–21.
24. Zhang, T. and Zhu, Q. (2018). Distributed privacy-preserving collaborative intrusion detection systems for VANETs. *IEEE Transactions on Signal and Information Processing over Networks, 4*(1), pp. 148–161.
25. Kumar, N. and Chilamkurti, N. (2014). Collaborative trust aware intelligent intrusion detection in VANETs. *Computers & Electrical Engineering, 40*(6), pp. 1981–1996.
26. Lal, A. S. and Nair, R. (2015). Region authority based collaborative scheme to detect Sybil attacks in VANET. In 2015 International Conference on Control Communication & Computing India (ICCC) (pp. 664–668). IEEE.

27. Zhang, J. (2011). A survey on trust management for vanets. In 2011 IEEE International Conference on Advanced Information Networking and Applications (pp. 105–112). IEEE.
28. Mejri, M. N., Ben-Othman, J., and Hamdi, M. (2014). Survey on VANET security challenges and possible cryptographic solutions. *Vehicular Communications*, *1*(2), pp. 53–66.
29. Engoulou, R. G., Bellaïche, M., Pierre, S., and Quintero, A. (2014). VANET security surveys. *Computer Communications*, *44*, pp. 1–13.
30. Yan, G., Olariu, S., and Weigle, M. C. (2008). Providing VANET security through active position detection. *Computer Communications*, *31*(12), pp. 2883–2897.
31. Todorova, M. S. and Todorova, S. T. (2016). DDoS attack detection in SDN-based VANET architectures. *No. June, 175*.
32. Bhoi, S. K. and Khilar, P. M. (2013). A secure routing protocol for Vehicular Ad Hoc Network to provide ITS services. In 2013 International Conference on Communication and Signal Processing (pp. 1170–1174). IEEE.
33. Bhoi, S. K. and Khilar, P. M. (2014). SIR: A secure and intelligent routing protocol for vehicular ad hoc network. *IET Networks*, *4*(3), pp. 185–194.
34. Bhoi, S. K. and Khilar, P. M. (2016). RVCloud: A routing protocol for vehicular ad hoc network in city environment using cloud computing. *Wireless Networks*, *22*(4), pp. 1329–1341.
35. Swain, R. R., Khilar, P. M., and Bhoi, S. K. (2018). Heterogeneous fault diagnosis for wireless sensor networks. *Ad Hoc Networks*, *69*, pp. 15–37.
36. Bhoi, S. K., and Khilar, P. M. (2012). SST: A secure fault-tolerant Smart Transportation system for Vehicular Ad hoc Network. In 2012 2nd IEEE International Conference on Parallel, Distributed and Grid Computing (pp. 545–550). IEEE.
37. Bhoi, S. K. (2013). SGIRP: A Secure and Greedy Intersection-Based Routing Protocol for VANET using Guarding Nodes (Doctoral dissertation).
38. Bhoi, S. K. and Khilar, P. M. (2016). Self soft fault detection based routing protocol for vehicular ad hoc network in city environment. *Wireless Networks*, *22*(1), pp. 285–305.
39. Bhoi, S. K., Nayak, R. P., Dash, D., and Rout, J. P. (2013). RRP: A robust routing protocol for Vehicular Ad Hoc Network against hole generation attack. In 2013 International Conference on Communication and Signal Processing (pp. 1175–1179). IEEE.
40. Nayak, R. P., Sethi, S., and Bhoi, S. K. (2018). PHVA: A position based high speed vehicle setection algorithm for detecting high speed vehicles using vehicular cloud. In 2018 International Conference on Information Technology (ICIT) (pp. 227–232). IEEE.
41. Varga, A., and Hornig, R. (2008). An overview of the OMNeT++ simulation environment. In Proceedings of the 1st international conference on Simulation tools and techniques for communications, networks and systems & workshops (p. 60). ICST (Institute for Computer Sciences, Social-Informatics and Telecommunications Engineering).

Part III

Applications

12 Cognitive Intelligence-Based Framework for Financial Forecasting

Sarat Chandra Nayak[1], Sanjib Kumar Nayak[2], Sanjaya Kumar Panda[3], and Ch. Sanjeev Kumar Dash[4]

[1]Department of Computer Science and Engineering, CMR College of Engineering & Technology, Hyderabad, India
[2]Department of Computer Application, VSS University of Technology, Odisha, India
[3]Department of Computer Science and Engineering, National Institute of Technology, Warangal, India
[4]Department of Computer Science and Engineering, Silicon Institute of Technology, Bhubaneswar, India

1 INTRODUCTION

Financial time series (FTS) is characteristically random when compared with the general time series. It is characterized by factors such as high non-uniformity, dynamicity, and turbulence. It is essential to perform frequent price monitoring of stock market indexes to study or witness the future price values of FTS data. Common FTS data include indexes of daily stock index closing, foreign exchange training, and price of crude oil. Forecasting of FTS data is the method of predicting the financial performance of a stock market in the future based on current or past stock market index values. The behavioral pattern of FTS typically resembles a process with an arbitrary walk. Due to its high degree of uncertainty, FTS forecasting is often regarded as an overly complex and critical task. FTS shows arbitrary fluctuations as they are related to various socio-economic factors directly or indirectly. They are highly nonlinear and volatile in nature. Thus, it is difficult to predict the FTS. Computational intelligent methods have proven their superiority over statistical methods in dealing with the nonlinearity of FTS [1–5]. Artificial neural networks (ANNs) are pattern recognition algorithms in complex and nonlinear datasets. ANNs are a special category of cognitive intelligence-based methods and are recognized as universal approximation functions. The ANNs are nonlinear, data-driven, self-adaptive in nature, and can work on a few apriori assumptions. The high accuracy of ANN-based models over linear and naïve methods for FTS prediction is established in many research [6–8]. Apart from these, several models based on ANN and their hybridization are suggested in literature and claimed to be better than traditional methods [9–14]. ANN and its combination with linear methods are producing better forecasting accuracy than others are suggested by few researchers [15,16].

Training of ANN plays a vital role in its applicability. Gradient descent-based algorithms are commonly adopted to train ANN models. However, swarm and evolutionary-based optimization techniques have been proposed in the last decade and mostly used for ANN training, which performed better than gradient descent-based methods in terms of landing at global minima with faster convergence. The widely used swarm intelligence methods include particle swarm optimization [17], artificial bee colony optimization [18], ant colony optimization [19], and so on. Based on the idea of the natural evolution process, several evolutionary algorithms are proposed, including a genetic algorithm [20], differential evolution [21], and so on. Few methods are framed by following the migration behavior of birds and animals such as animal migration optimization [22] and monarch butterfly optimization (MBO) [23]. One optimization technique is based on the explosion process of fireworks termed as fireworks algorithm [24]. A recently proposed metaheuristic termed as Multi-Verse Optimizer (MVO) [25] is derived from the multi-universe existence hypothesis and their interactions through white-, black-, and wormholes. Moreover, the global optima can be found by the population-based optimization method.

The intention here is to design a cognitive intelligence-based framework using ANN. To overcome the limitations of traditional backpropagation-based training, we discuss here a few recently introduced metaheuristics for the training of the artificial neural network. Metaheuristic optimization techniques include methods like GA, MVO, MBO, DE, FWA, and PSO to find the optimal ANN model and its parameters. The model of an optimal ANN is then deployed for modeling and forecasting of a typical financial time series forecasting application consisting of factors like daily stock market closing indexes, exchange rates, and crude oil prices.

Few major contributions of the chapter are:

- Exploring the application of the ANN for FTS.
- Exploring metaheuristics-based optimization techniques for artificial neural network training.
- Suitability forecasting of ANN models by the systematic experimental result analysis.

The remaining part of the chapter is organized in the following manner. A brief insight into ANN is given in Section 2. ANN training methods are discussed in Section 3. FTS data are discussed in Section 4. Section 5 focuses on ANN-based FTS forecasting. Simulation outputs are summarized in Section 6. Section 7 represents the conclusion and future research directions.

2 ARTIFICIAL NEURAL NETWORK

In complex and non-linear datasets, ANN helps discover hidden patterns. The basic computing element in ANN is a neuron. Unlike biological neurons, it can perceive the environment, compute, and store the computed sum for future processes. The whole network is layer-based and made of nodes. Computation is performed by the nodes by combining input weighted data that either amplify or reduce input levels. Summation of input-weight products is performed and passed through an activation function, to find whether and to what extent that signal affects the outcome. The network is a fully connected one. The computed signals are passed from one layer to another and finally reach the output neuron. The intermediate layers are termed as hidden layers. A typical ANN architecture is shown in Figure 12.1. For simplicity, we presented here a network with only one hidden layer.

As depicted, the input layer contains n neurons to get the input signals. Every neuron is interconnected to other hidden neurons. An input-hidden layer link is formed for information transfer between the input and hidden layers. Each link is associated with a weight value that decides the extent of the input signal to be transferred. At each hidden layer neuron, computation is carried out followed by activation. In the same way, the computed sum is transferred to the output layer. This single output unit of the model estimates FTS data points using supervised learning. The linear transfer function is used by input neurons; hidden and output neurons use activations like *tanh*, *ReLu*, *sigmoid* function, etc. as follows.

$$y_{out} = 1/1 + e^{-y_{in}} \qquad (12.1)$$

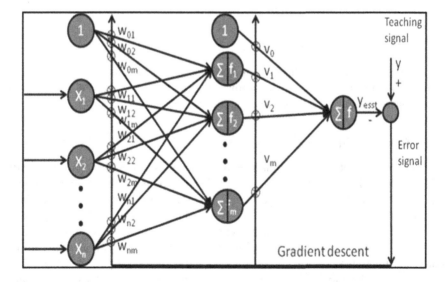

FIGURE 12.1 A Single Hidden-Layer ANN-Based Forecast.

where
    y_{out} = neuron output,
    y_{in} = neuron input.

The foremost layer nodes correspond to input variables with one node for each. The hidden neurons explore the non-linear relations among variables. The output at each hidden neuron is calculated as:

$$y_H = f(bias_j + \sum_{i=1}^{n} w_{i,j} * x_i) \qquad (12.2)$$

where x_i is the ith component of the input vector, w_{ij} is the synaptic weight between i^{th} input neuron and j^{th} hidden neuron, and $bias_j$ is the bias value. The model estimation (i.e., y_{ess}) is calculated using Equation (12.3).

$$y_{esst} = f(bias_0 + \sum_{j=1}^{m} v_j * y_H) \qquad (12.3)$$

where ... is the synaptic weight value from j^{th} hidden neuron to output neuron, y_H is the weighted sum calculated as in Equation (12.2), and $bias_0$ is the output bias. The deviation is calculated as:

$$error_i = |y_i - y_{esst}| \qquad (12.4)$$

where $error_i$ the error signal, y_i is the actual for i^{th} training sample. In training, the model is iteratively trained with the training vectors. The weight and biases are optimized by a learning algorithm until the desired input-output mapping occurs. The aim is minimization of error through optimization of weight and bias. Different training methods are discussed in the next section.

3 ANN TRAINING METHODS

As mentioned earlier, the efficiency of ANN-based models relies heavily on model training. We present here two well-known training methods: gradient descent and evolutionary optimization-based training algorithms.

3.1 Gradient Descent-Based Method

Gradient-based methods for minimization of errors train backpropagation (BP) ANN (BPNN). BP is a supervised learning algorithm that estimates the error in the output by finding the gradient of generated error and performing the weight and bias adjustment of ANN in the direction of the descending gradient. BPNN is a popular in-stock prediction with improved accuracy, dispersal abilities, and enhanced nonlinearity mapping. Gradient descent focuses on weight update in ANN; it updates and tunes the ANN parameters to minimize the loss function. BPNN first propagates in the forward direction by computing the dot product of inputs and corresponding weights. The activation function is then applied to the weighted sum

Cognitive Intelligence-Based Framework

that transforms it into an output signal. It introduces nonlinearities to the model and learns any arbitrary functional mappings. Afterward, it propagates backward direction carrying the error terms and uses a gradient descent approach for weight update. Corresponding to the weights, the error function gradient is calculated. It then updates the parameters in the reverse direction of the cost function gradient w. r. t to the model parameters. The major shortcomings of backpropagation neural networks are the local minima finding, large randomness, etc. that impacts the stock price prediction accuracy. These disadvantages encouraged the development of evolutionary optimization-based learning algorithms by merging nonlinear models (ANN, etc.), evolutionary soft computing techniques (GA, etc.), and hybrid search techniques for improved performance.

3.2 Evolutionary Optimization-Based Method

Evolutionary optimization algorithms are motivated by biological/natural evolution. They are also called nature-inspired computing methodologies as they mimic the behavior of nature. These are established as more efficient in searching global optima than traditional gradient descent-based techniques. The major steps in almost nature-inspired computing methods are as follows:

- A new search space can be created out of the existing one.
- The optimality of a solution can be evaluated by measuring its fitness.
- Design variables are optimized through the application of evolutionary search operators.
- It can reach a global optimum by using random search techniques.

For the sake of space, we present here only three recently introduced evolutionary algorithms such as FWA, MVO, and MBO. Later, the performance of these will be compared with quite common popular methods like GA, DE, and PSO.

3.2.1 FWA

Fireworks algorithm (FWA) is a novel optimization approach to simulate fireworks explosion [23]. In a search space, FWA aims to select multiple locations to generate a set of sparks by a fireworks explosion. For next-generation, qualitative firework locations are selected. The iteration process continues until the optimum value or the stopping criterion. This approach is a three-step process: (1) perform N fireworks at N chosen places, (2) find out the locations of sparkle on blast and evaluate, and (3) end upon the attainment of optimal location or decide on N new locations for the subsequent generation of the explosion. Fireworks explosion resembles exploration practice in the neighborhood. Here, for each firework x_i, the amplitude of flare-up (A_i) and number of sparks (s_i) are as follows:

$$A_i = \hat{A} \cdot \frac{f(x_i) - f_{min} + \varepsilon}{\sum_{j=1}^{p}(f(x_j) - f_{min}) + \varepsilon} \tag{12.5}$$

$$s_i = \frac{m \cdot f_{max} - f(x_i) + \varepsilon}{\sum_{j=1}^{p} (f_{max} - f(x_j)) + \varepsilon}, \qquad (12.6)$$

where $\hat{A}$ represents maximum value of explosion amplitude. f_{max} and f_{min} represent the maximum and minimum value of objective function within p fireworks. m represents a parameter to control the number of sparks generated, and ε is a constant to stay away from zero division mistake. To prevail over the negative effects, bounds are imposed on s_i as follows:

$$s_i = \begin{cases} s_{max}, & \text{if } s_i > s_{max} \\ s_{min}, & \text{if } s_i < s_{min} \\ s_i, & \text{otherwise} \end{cases} \qquad (12.7)$$

Each spark location x_j generated by x_i is considered by setting z-directions randomly and each respective dimension k for setting the component x_j^k based on x_i^k, where $1 \leq j \leq s_i$, $1 \leq k \leq z$.

The setting of x_j^k follows a two-way approach:

- For most sparks, add displacement to x_j^k as:

$$x_j^k = x_i^k + A_i \cdot rand(-1, 1) \qquad (12.8)$$

- To preserve heterogeneity, for few specific sparks, a Gaussian distribution-based explosion coefficient is applied to x_j^k as:

$$x_j^k = x_i^k \cdot Gaussian(1, 1) \qquad (12.9)$$

A new location beyond the search space mapped into as follows:

$$x_j^k = x_{min}^k + |x_j^k| \% (x_{max}^k - x_{min}^k), \qquad (12.10)$$

In the next step N other locations are selected keeping the current best location $x*$. The remaining locations are measured based on their detachment from others computed as in Equation (12.11).

$$Distance(x_i) = \sum_{j \in K} \|x_i - x_j\|, \qquad (12.11)$$

A location x_i is selected based on a probability value for the next generation as follows:

$$prob(x_i) = \frac{Distance(x_i)}{\sum_{j \in K} Distance(x_j)} \qquad (12.12)$$

FWA fundamentals are designed and depicted below.

Algorithm 1: FWA

Pick N locations randomly for fireworks;
   While (stopping criteria not met)
   Set out N fireworks at N locations
   For firework x_i
   Calculate s_i using (6)
   Get locations of s_i sparks of firework x_i using (7)
   End
   For k = 1 to m
   Select x_j randomly
   Generate a spark using (8)
   End
   Select x^* and carry on it next generation
   Select *N-1* locations randomly based on (12)
End

As the explosion power affects the sparks, it results in simultaneous *z-direction* movements. Therefore, FWA converges faster. Generation of spark and the selection of specific locations prevent early convergence of FWA [23]. The article embodies the advantages of FWA over other optimization algorithms [23].

3.2.5 MBO

MBO is a recently proposed nature-inspired optimization algorithm [24]. This is inspired by the concept of monarch butterfly migration between regions of Canada, the USA, and Mexico. The former location is considered as Land1 (subpopulation1), and the latter is Land2 (subpopulation2). Individuals of monarch butterflies are in either of the two distinctive lands. The offspring are generated by the application of migration operator on the butterflies in Land1 or Land2. As per the natural concept, the monarch butterflies stay approximately five months in Land1 and seven months in Land2. Therefore, the population size of Land1 can be considered as $NP_1 = ceil\ (p * NP)$ and that of Land2, $NP_2 = NP - NP_1$, where p represents the ratio of butterflies in Land1 and *NP* is the overall population size. A new individual is formed from the old one with the migration operator. The *kth* element of butterflies x_i in *t+1* generation can be represented as follows.

$$\begin{cases} x_{i,k}^{t+1} = x_{r1,k}^{t}, & if\ r \le p \\ x_{i,k}^{t+1} = x_{r2,k}^{t}, & if\ r > p \end{cases} \quad (12.13)$$

where $r1$ and $r2$ are two randomly selected monarch butterflies from Land1 and Land 2. t represents the current generation and r is calculated as:

$$r = rand * migration\ period \quad (12.14)$$

where rand follows uniform distribution for randomness, and the *migration period* is selected as 1.2 (MBO algorithm basics).

As per the above discussion, the bigger the value of p, the more are the samples selected from Land1 (subpopulation1); otherwise, more samples are selected from subpopulation2. Regarding p-value, a subpopulation is regarded as a key component. By adjusting the value of p, the direction of migration operation can be balanced.

The next operator (i.e. butterfly adjusting operator) is used to update the butterfly position. Here, k^{th} element of jth butterfly at iteration $t + 1$ can be updated as follows.

$$\begin{cases} x_{j,k}^{t+1} = x_{best,k}^{t}, & if\ r \le p \\ x_{j,k}^{t+1} = x_{r3,k}^{t}, & if\ r > p \end{cases} \quad (12.15)$$

where x_{best}^{t} the best monarch butterfly of the population is so far obtained, x_{r3}^{t} is randomly selected for subpopulation2 and t is the current generation. To accomplish further searching, it can be updated as follows.

$$x_{j,k}^{t+1} = x_{j,k}^{t} + \alpha * (dx_k - 0.5),\ if\ rand > Butterfly\ adjusting\ rate \quad (12.16)$$

where
α = weighting factor and
dx = walk-step of butterfly.
The parameter dx and α can be computed as follows.

$$dx = Levy(x_j^t) \quad (12.17)$$

$$\alpha = \frac{S_{max}}{t^2} \quad (12.18)$$

The bigger value of α increases exploration and the smaller value encourages search space exploration. S_{max} is the maximum walk step for a one-step butterfly movement. As per the discussion above, the MBO process is framed as follows.

Algorithm 2 MBO algorithm

1. **Initialization**
   Initialize population P randomly
   Set counter t = 1 and Maximum generation
   Set size of NP1 and NP2
   Set S_{max}, BAR, *migration period*, and migration ratio p
2. **Evaluation**
   Evaluate the fitness of all butterflies according to their position
3. ***While*** (t < MaxGen)
   Sort the butterflies on their fitness
   Divide all butterflies into Land1 and Land2
   for I = 1: NP1
   Generate Offspring using Equation (12.1) and add to subpopulation1
   end for
   for j = 1: NP2
   Generate Offspring using Equations (12.13) and (12.5), add to sub-population2
   end for
   New population = Subpopulation1 U Subpopulation2
   Evaluate new population
   $t = t+1$
   end while
4. **Output** best butterfly as best solution

3.2.6 MVO

The process of MVO is initiated with a random solution set which is periodically updated with finite iterations. In MVO, the jth parameter of ith solution (w_i^j) is updated as:

$$w_i^j = \begin{cases} \begin{cases} w_j + TDR + ((ub_j - lb_j) * rand_3 + lb_j), if rand_2 < 0.5 \\ if rand_1 < WEP \\ w_j - TDR + ((ub_j - lb_j) * rand_3 + lb_j), if rand_2 \geq 0.5 \end{cases} \\ W_{Roulettewheel}^j \, if rand_1 \geq WEP \end{cases} (1) \quad (12.19)$$

where
$rand_1$, $rand_2$ and $rand_3$ are random numbers in [0, 1],
$W_j = j^{th}$ parameter of best weight set,
ub_j = Upper bound of jth component,
lb_j = Lower bound of jth component,
$w_{Roulettewheel}^j = j^{th}$ part of a solution selected by the roulette wheel.

The Wormhole Existence Probability is calculated as:

$$WEP = \min + current_{iter} * \left(\frac{\max - \min}{Max_{iteration}}\right) \qquad (12.20)$$

TDR = travelling distance rate coefficient, calculated as:

$$TDR = 1 - \left(\frac{current_{iter}}{Max_{iteration}}\right)^{\frac{1}{p}} \qquad (12.21)$$

In the process of optimization, factors such as WEP and TDR judge the frequency and the extent of solution change. Exploitation accuracy is denoted by parameter p. Enhancement of WEP value is directly proportional to the rate of exploitation. Exploitation is obtained by substituting the jth element of the solution with the solution using the roulette wheel method. It is beneficial to avoid the local optimum value. Here, the present solution has a black hole, and the best solution has a white hole. In the optimization process, WEP and TDR may be adaptively changed to investigate exploration and exploitation. The current best solution chooses white holes with a roulette wheel (in proportion to relative fitness costs), improves the less accurate solutions of the search process, and generates black holes negatively correlated to the fitness values.

4 FINANCIAL TIME SERIES DATA

Three types of FTS data are considered here: (1) every day closing prices of stock exchanges, (2) exchange rate pattern, and (3) daily and weekly crude oil prices series. As said above, these FTS exhibits characteristics of high volatility, nonlinearity, and chaos. The FTS data, category, sources, and period of the collection are represented in Table 12.1. We then analyzed the statistical aspects of attributes of all FTS; the outcomes are depicted in Table 12.2. Usually, FTS data show randomness and possesses no serial correlation and heteroskedasticity. We tested all FTS for randomness using Box and Pierce test method and no serial correlation using Durbin and Watson test method to ensure randomness. The computed values are summarized in Table 12.3. These evidences are in support of the autocorrelation function and partial autocorrelation function test.

5 ANN-BASED FINANCIAL PREDICTION

Conventional time-series can be represented by a discrete sequence represented as $\{x_1, x_2, x_3, \cdots, x_n\}$, where n represents count of observations, x represents discrete data points occurring at equal time intervals. An FTS with deterministic and stochastic components as interference from noise can be represented as:

$$p_{t+1} = f(p_t, p_{t-1}, \cdots, p_{t-n}), \qquad (12.22)$$

The sliding window method generates input training and test from the original FTS [5]. For example, the generated sample training and sample test patterns with sliding window length three are shown as follows:

TABLE 12.1
FTS Information

FTS	Stock Market Name	Duration	Source
Closing prices series	BSE NASDAQ DJIA FTSE	02 January 2016 to 20 September 2019	https://in.finance.yahoo.com/quote/
Exchange rates series	INR/USD CNY/USD PKR/USD BDT/USD	07 February 2019 to 05 August 2019	https://www.exchangerates.org.uk/
Crude oil prices series	Daily Crude Oil Prices (DCO) Weekly Crude Oil Prices (WCO)	April 1983 to July 2019	http://www.eia.doe.gov/

TABLE 12.2
FTS Statistics

FTS	Statistics with Description					
	Min.	Max.	Mean	Std Dev.	Skewness	Kurtosis
BSE	1.8297e+03	2.5980e+03	2.1605e+03	206.8555	0.4131	4.8323
NASDAQ	1.8321e+04	3.3152e+04	2.5132e+04	3.4674e+03	0.3928	5.5981
DJIA	1.2318e+04	3.0152e+03	1.8435e+04	3.0093e+03	0.3827	5.1087
FTSE	1.6225e+03	2.5106e+03	2.3362e+04	2.8495e+03	0.4211	4.5962
INR/USD	68.4103	71.3828	69.6291	0.7640	0.5081	5.3777
CNY/USD	6.6872	7.0497	6.8055	0.0883	1.9776	15.4469
PKR/USD	138.6879	163.5410	147.3221	8.2301	2.3403	16.6110
BDT/USD	83.6815	85.0451	84.2992	0.2609	-0.1626	5.9754
DCO	10.4200	145.2900	42.9828	28.5640	0.9953	2.8829
WCO	11.0900	142.4600	42.8949	28.5241	1.0011	2.8921

$$\begin{matrix} p(i) & p(i+1) & p(i+2) & \vdots & p(i+3) \\ p(i+1) & p(i+2) & p(i+3) & \vdots & p(i+4) \\ \underbrace{p(i+2) \quad p(i+3) \quad p(i+4)}_{\text{Training data}} & \vdots & \underbrace{p(i+5)}_{\text{Target}} \end{matrix}$$

$$\underbrace{p(i+3) \quad p(i+4) \quad p(i+5)}_{\text{Test data}} \vdots \underbrace{p(i+6)}_{\text{Target}}$$

Sigmoid normalization is used to normalize different training and test patterns [26] as shown in Equation (12.23).

TABLE 12.3
Computed *p*-Values from Box and Pierce Test and Durbin Watson Test

FTS	Box and Pierce	Durbin and Watson
BSE	0.9868	1.7447e-11
NASDAQ	0.0125	9.3620e-12
DJIA	8.0381	1.9664e-11
FTSE	0.9489	9.4633e-12
INR/USD	0.8247	1.7286e-11
CNY/USD	0.2418	1.9263e-10
PKR/USD	0.2192	8.2635e-11
BDT/USD	0.0012	6.2665e-10
DCO	0.6412	1.7663e-10
WCO	0.0040	9.2273e-12

$$p_{norm} = \frac{1}{1 + e^{-\left(p_i - p_{min}/p_{max} - p_{min}\right)}} \quad (12.23)$$

where

p_{norm} = normalized price,
p_i = current day closing price,

p_{max} and p_{min} are the high and low price of the window, respectively. Current training set data are normalized within [0, 1].

Seven different models are developed in this work for one-day-ahead forecasting of ten FTS. The ANN architecture shown in Figure 12.1 is used as the base model. Seven different training algorithms are used to train the ANN separately. The model names are shown in Table 12.4. The overall steps for ANN-based forecasting are shown in Figure 12.2.

6 SIMULATION STUDIES AND RESULTS ANALYSIS

Separate experiments were conducted for seven models using ten different FTS. The training and test samples were the same for all models. The normalized mean squared error (NMSE), as shown in Equation (12.20), was considered as the performance metric (Table 12.5).

$$NMSE = \frac{1}{N} \sum_{i=1}^{N} (x_i - \hat{x}_i)^2 \quad (12.24)$$

In the above equations:
x_i = closing price (exact)
$\hat{x}_i$ = price (estimated)

TABLE 12.4
Models Developed

S. No.	Model Combined with Artificial Neural Network	Short Name
1	BP	BPNN
2	PSO	PSO-ANN
3	GA	GA-ANN
4	DE	DE-ANN
5	MBO	MBO-ANN
6	MVO	MVO-ANN
7	FWA	FWA-ANN

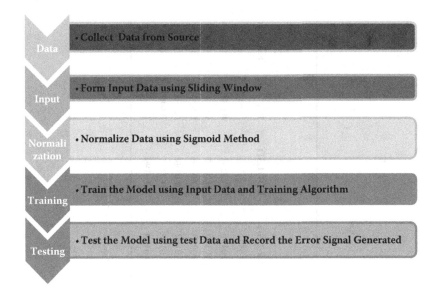

FIGURE 12.2 Overall Steps in ANN-Based Forecasting.

N is the quantity of training sample in the dataset (total)

The NMSE values get better as they near zero. For each model, the average of 20 simulations was recorded and Table 12.5 summarizes the details. The values of NMSE in bold represent the best values. Results conclude that the hybrid models perform quite better than the BPNN model. Comparing evolutionary-based models, PSO-ANN achieved three times the best NMSE value, and GA-ANN got two times the best value. The MBO-ANN achieved the best NMSE value the highest number of times (i.e. five times). FWA-ANN got the best value twice, followed by MVO-ANN once. These new algorithms (i.e. MVO, FWA, and MBO) are found quite competitive to other popular methods (GA, DE, and PSO). The computation times are summarized in Table 12.6.

TABLE 12.5
NMSE from All Models

Model	INR/USD	CNY/USD	PKR/USD	BDT/USD	BSE	NASDAQ	DJIA	FTSE	DCO	WCO
BPNN	0.0876	0.0653	0.0537	0.0750	0.0592	0.0257	0.0382	0.0490	0.0487	0.0436
PSO-ANN	0.0255	0.0169	0.0195	0.0247	**0.0289**	0.0183	**0.0270**	0.0346	**0.0163**	0.0278
GA-ANN	**0.0173**	0.0176	0.0198	0.0255	0.0334	0.0175	**0.0270**	0.0373	0.0166	0.0277
DE-ANN	0.0276	0.0188	0.0213	0.0236	0.0317	0.0185	0.0289	0.0375	0.0213	0.0277
MBO-ANN	0.0178	0.0158	**0.0192**	**0.0194**	0.0305	0.0174	0.0248	**0.0338**	**0.0163**	**0.0253**
MVO-ANN	0.0176	0.0183	0.0198	0.0233	0.0401	**0.0167**	0.0355	0.0409	0.0167	0.0264
FWA-ANN	0.0182	**0.0155**	0.0199	0.0199	0.0308	0.0176	0.0272	**0.0338**	0.0179	0.0278

TABLE 12.6
Computation Time from All Models

Model	FTS INR/USD	PKR/USD	BDT/USD	CNY/USD	BSE	NASDAQ	DJIA	FTSE	DCO	WCO
BPNN	65.98	64.85	66.75	73.25	103.55	98.29	101.20	113.00	93.25	65.75
PSO-ANN	90.75	83.65	86.34	90.44	100.35	116.05	114.35	120.35	88.65	80.35
GA-ANN	94.15	89.88	88.95	93.18	114.26	120.85	121.85	124.20	90.18	83.95
DE-ANN	85.30	83.35	84.77	91.45	100.27	117.70	112.08	110.22	90.45	80.07
MBO-ANN	77.46	74.47	84.40	92.75	109.45	109.06	99.846	107.43	88.75	81.42
MVO-ANN	82.04	73.95	73.87	86.99	103.35	109.85	105.85	115.35	85.74	83.80
FWA-ANN	78.48	79.50	85.55	93.25	100.65	113.60	110.95	120.35	90.22	85.05

The evolutionary optimization-based models are a little higher computation time than BPNN. To support the efficiency of evolutionary optimization-based ANN models over the gradient descent-based method, we calculated the percentage of reduction in NMSE as in Equation (12.25) and shown in Figure 12.3.

$$Reduction\ in\ NMSE = \frac{(NMSE\ of\ BPNN - NMSE\ of\ a\ hybrid\ model)}{NMSE\ of\ BPNN} * 100$$

(12.25)

Based on the above analysis, the major observations can be summarized as follows.

- ANN-based forecasting models can capture the non-linearities of FTS
- Evolutionary optimization-based ANN models perform better than traditional gradient descent-based ANN
- The recently proposed evolutionary optimization algorithms are performing at par —sometimes better — with quite common evolutionary optimization techniques.

7 CONCLUSIONS

The high impact of unpredictability makes FTS forecasting a challenging task. FTS shows arbitrary fluctuations as they talk with various socio-economic factors directly or indirectly. This chapter discussed the use of one of the powerful cognitive intelligence

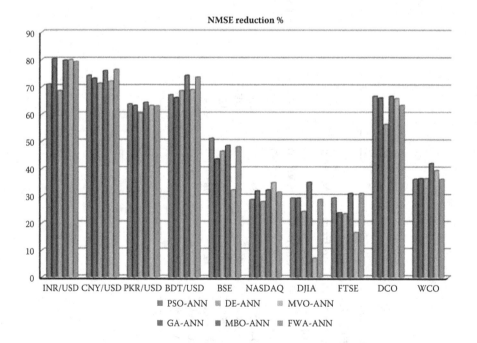

FIGURE 12.3 NMSE Reduction on Adopting Evolutionary Optimization-Based Model.

methods (i.e. ANN applications for FTS forecasting). To avoid the limitations of traditional backpropagation-based training, three recently introduced evolutionary optimization methods such as MBO, FWA, and MVO are explored as a training algorithm for ANN. The comparative study included other hybrid models developed — GA, DE, and PSO. The optimal ANN structure is then deployed to model and forecast ten real FTS. These new methods are found competitive and at par with the popular GA-, DE-, and PSO-based ANN models. Particularly, the MBO-ANN-based model is found as the most efficient. These models may be applied to other kinds of data mining problems. Variants of these methods may be explored.

REFERENCES

1. Hsu, M. W., Lessmann, S., Sung, M. C., Ma, T., and Johnson, J. E. (2016). Bridging the divide in financial market forecasting: machine learners vs. financial economists. *Expert Systems with Applications*, 61, 215–234.
2. Adhikari, R. and Agrawal, R. K. (2014). A combination of artificial neural network and random walk models for financial time series forecasting. *Neural Computing and Applications*, 24(6), 1441–1449.
3. Zhang, G. P. (2003). Time series forecasting using a hybrid ARIMA and neural network model. *Neurocomputing*, 50, 159–175.
4. Zhang, H., Kou, G., and Peng, Y. (2019). Soft consensus cost models for group decision making and economic interpretations. *European Journal of Operational Research*, 277(3), 964–980.
5. Nayak, S. C., Misra, B. B., and Behera, H. S. (2019). ACFLN: Artificial chemical functional link network for prediction of stock market index. *Evolving Systems*, 10(4), 567–592.
6. Qiu, M., Song, Y., and Akagi, F. (2016). Application of artificial neural network for the prediction of stock market returns: The case of the Japanese stock market. *Chaos, Solitons & Fractals*, 85, 1–7.
7. Aras, S. and Kocakoç, İ. D. (2016). A new model selection strategy in time series forecasting with artificial neural networks: IHTS. *Neurocomputing*, 174, 974–987.
8. Mostafa, M. M. (2010). Forecasting stock exchange movements using neural networks: Empirical evidence from Kuwait. *Expert Systems with Applications*, 37(9), 6302–6309.
9. Guresen, E., Kayakutlu, G., and Daim, T. U. (2011). Using artificial neural network models in stock market index prediction. *Expert Systems with Applications*, 38(8), 10389–10397.
10. Lu, C. J. and Wu, J. Y. (2011). An efficient CMAC neural network for stock index forecasting. *Expert Systems with Applications*, 38(12), 15194–15201.
11. Nayak, S. C. and Misra, B. B. (2018). Estimating stock closing indices using a GA-weighted condensed polynomial neural network. *Financial Innovation*, 4(1), 21.
12. Zhong, X. and Enke, D. (2019). Predicting the daily return direction of the stock market using hybrid machine learning algorithms. *Financial Innovation*, 5(1), 4.
13. Nayak, S. C., Misra, B. B., and Behera, H. S. (2018). On developing and performance evaluation of adaptive second order neural network with ga-based training (asonn-ga) for financial time series prediction. In *Advancements in Applied Metaheuristic Computing* (pp. 231–263). IGI global.
14. Nayak, S. C., Misra, B. B., and Behera, H. S. (2017). Exploration and incorporation of virtual data positions for efficient forecasting of financial time series. *International Journal of Industrial and Systems Engineering*, 26(1), 42–62.

15. Khandelwal, I., Adhikari, R., and Verma, G. (2015). Time series forecasting using hybrid ARIMA and ANN models based on DWT decomposition. *Procedia Computer Science*, *48*(1), 173–179.
16. Mallikarjuna, M., Guptha, K. S., and Rao, R. P. (2017). Modelling sectoral volatility of Indian stock markets. *Wealth International Journal of Money Banking and Finance*, *6*(2), 4–9.
17. Kennedy, J. and Eberhart, R. (1995, November). Particle swarm optimization. In Proceedings of ICNN'95-International Conference on Neural Networks (Vol. 4, pp. 1942–1948). IEEE.
18. Karaboga, D. and Basturk, B. (2007). A powerful and efficient algorithm for numerical function optimization: Artificial bee colony (ABC) algorithm. *Journal of Global Optimization*, *39*(3), 459–471.
19. Dorigo, M., Maniezzo, V., and Colorni, A. (1996). Ant system: Optimization by a colony of cooperating agents. *IEEE Transactions on Systems, Man, and Cybernetics, Part B (Cybernetics)*, *26*(1), 29–41.
20. Goldberg, D. E. (1989). Genetic algorithms in search, optimization, and machine learning, Addison-Wesley, Reading, MA, 1989. *NN Schraudolph and Journal*, *3*(1).
21. Storn, R. and Price, K. (1997). Differential evolution–a simple and efficient heuristic for global optimization over continuous spaces. *Journal of Global Optimization*, *11*(4), 341–359.
22. Li, X., Zhang, J., and Yin, M. (2014). Animal migration optimization: an optimization algorithm inspired by animal migration behavior. *Neural Computing and Applications*, *24*(7-8), 1867–1877.
23. Tan, Y. (2015). *Fireworks Algorithm* (pp. 17–35). Heidelberg: Springer.
24. Wang, G. G., Deb, S., and Cui, Z. (2019). Monarch butterfly optimization. *Neural Computing and Applications*, *31*(7), 1995–2014.
25. Mirjalili, S., Mirjalili, S. M., and Hatamlou, A. (2016). Multi-verse optimizer: A nature-inspired algorithm for global optimization. *Neural Computing and Applications*, *27*(2), 495–513.
26. Nayak, S. C., Misra, B. B., and Behera, H. S. (2014). Impact of data normalization on stock index forecasting. *International Journal of. Computer Information Systems and Industrial Management Applications*, *6*, 357–369.

13 Benefits of IoT in Monitoring and Regulation of Power Sector

Harpreet Kaur Channi
Assistant Professor, EE Department, UIE,
Chandigarh University, Gharuan, Mohali, India

1 INTRODUCTION

The entire installed ability of India's power distribution is over 350 GW as of August 2019 [1]. Renewable power, including large hydro generation, constitutes almost 35% of this capacity. This makes India the third chief energy producer and third main energy consumer worldwide. The installed non-conventional energy capacity of India is 80.5 GW as of June 2019. The MNRE (Ministry of New and Renewable Energy) has targeted to achieve a total potential of 175 GW from renewable energy sources by December 2022 [1]. To manage the humongous power distribution network, the Indian government has embarked on several policy initiatives to implement far-reaching improvement in the distribution area, through the administration of modern technology initiatives.

1.1 Indian Power Distribution Reforms Scenario

Following the R-APDRP (Restructured Accelerated Power Development and Reform Program) to enable IT and automate the power distribution systems, another scheme called IPDS (Integrated Power Development Scheme) has been initiated by the Ministry of Power to enable the Indian Power Distribution Companies (DISCOMs) and strengthen their transmission and distribution networks and subparts, implement automated reading of substation feeders, transformers (distribution type) and users, and continue sustained automation of power distribution management functions [2].

IPDS lays a lot of emphasis on 100% metering of power distribution supply sources and consumer off-take points, substation feeder monitoring, asset and fault management, and measurement of reliability and power quality indices (SAIDI, SAIFI, CAIDI, CAIFI, voltage quality index, THD, etc.), as per the Central Electricity Authority (CEA) guidelines [1,2]. The volatile renewable energy connections cause grid disturbances resulting from voltage sags, swells, transients, and harmonics.

To cope with these challenges, a model power quality regulatory framework has been introduced by CERC (Central Electricity Regulatory Commission) to assess and mitigate power quality issues like total harmonic distortion (THD), reactive power, and voltage quality index, in compliance with IEEE 519 and EN 50160 standards [2,3].

1.2 Emergence of Internet of Things (IoT)

Driven by an increasingly volatile electricity supply scenario, regulatory guidelines, and market forces, amid India's growing electricity demand, efficiency holds the key to sustainable operations. This realization has led to the gradual adoption of the latest expertise such as Big data, IoT (Internet of Things), cloud computing, and AI/ML in Indian DISCOMs [4]. IoT systems use sensors and internet technologies to capture real-time information of network assets and operation and transmit data to the distant control room for monitoring, visualization, and time-series analysis. Cloud computing, Big data, and AI/ ML technologies have been adopted by certain Indian utilities in power distribution management system (DMS) applications such as connections, metering and billing, work and asset management, GIS-based network analysis, energy audit, and customer services [5].

The increasingly digitalized world is rapidly becoming interconnected. Nearly every company will affect the Internet of Things (IoT) when computers start to interact and create decisions independently, without any human involvement [5,6]. Innovations extend from smart thermostats that optimize energy efficiency by changing the temperature of consumers' homes based on where they are, to refrigerators that order automatically when food runs low, and device component sensors that enable data collection, helping avoid expensive errors by promptly notifying maintenance requirements [7].

1.3 IoT in Renewable Energy Sources

IoT has enabled the creation of smart grids that support manual switching between renewables and long-established power plants to facilitate an uninterrupted power supply. This switching helps smart grids support the varying nature of renewable energy and facilitate non-stop energy supply to the consumers. Efficiency continues to be a significant barrier to the production of renewable energy [7]. This is especially true for methods that rely on variable resources such as wind and solar energy. For these electricity production methods, an intelligent IoT system enables the introduction of automated controls to increase performance. With the aid of these controls, it is possible to harvest energy from renewable resources with full output.

An IoT system can help predict the most optimal conditions for generating electricity. The equipment can be modified to produce optimum output as necessary. The biggest benefit of IoT systems is that they produce data in real-time, helping reduce wastage if any. IoT applications in renewable energy production include sensors that are connected to equipment for generation, transmission, and distribution. Such tools help businesses track and manage the equipment's running

Benefits of IoT in Monitoring

remotely in real-time [8]. This contributes to lower operating costs and reduces reliance on fossil fuels. Using renewable energy resources delivers a range of advantages over conventional ones. Implementation of IoT would help us make more use of these renewable energy sources.

2 IOT IN POWER SECTOR

Blackouts and load forecasting are two big problems in the power grid. During the transmission process in India, an additional 30% of energy vanishes. Fault conditions in the transmission network can result in blackouts of the power grid—often abrupt—and are difficult to find. Through load forecasting, one of IoT's most appealing applications is also expanded. All electronic unit (EU) utilities would therefore have smart meters nowadays to support this IoT-based smart load forecast [9,10].

IoT can play an important part in the construction of electrical systems that are extra powerful or "smart." The IoT is a pillar of "smart grids," which are basically a "data network that can intelligently integrate the actions of all related consumers—manufacturers, buyers, and those who do all—to effectively provide secure, affordable, and reliable sources of energy" [11]. Smart grid technologies include conveniently controllable two-way electrical power transfer, automatic, bidirectional knowledge transfer, as seen in Figure 13.1.

By fusing both—IoT and artificial learning—technologies will definitely help people overcome real-time problems. It will allow effective incorporation of infrastructure resources into communications and electrical power networks, and increase awareness about the power structure and the efficiency of infrastructure usage in the current power system [12]. As IoT technology is used in the smart grid, it can easily provide vital technological support—real-time control, maintenance assistance, generator fault location detection, transmission, substation, delivery, energy, and other power grid facets.

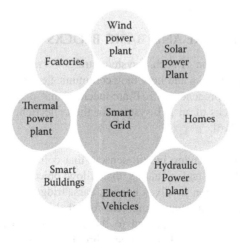

FIGURE 13.1 Smart Grid.

As the power sector is more competitive, smart technologies such as AI are required to efficiently monitor processes and extract value from all new data being produced. Using data analytics and machine learning, it is possible to solve various operational management challenges and create new concepts for medium and long-term planning—forecasting the availability of an asset, the importance of flexibility, and how to better use an asset to achieve optimum value. The electricity sector is experiencing a significant transition with the increased installation of green energy technology (solar photovoltaic and wind) providing unreliable energy supply, distributed energy resources (DERs), bidirectional electricity transmission, broad data flows produced by IoT and other tools, growing energy storage use, and shifting role for utilities and consumers.

While considering network decentralization through the adoption of distributed electricity generation and battery storage, IoT has tremendous potential for innovative management and business model approaches because of its data collection capability [13]. Implementing the unified energy network shifts to millions from providing hundreds of control points in a traditional power grid. Potential autonomous networks need micro-level supervision and control to attain their ability as utility providers and enhance the operation of electrical networks.

IoT technology can boost grid-connected smart asset efficiency, flexibility, and network operator control of these resources [14]. By linking energy suppliers, consumers, and grid infrastructure, IoT technology aims to encourage the growth of diverse networks and open new business possibilities by allowing customers to better leverage the value generated through their investments by providing specific services through demand-side management [7].

IoT is among the gravitating innovations and developments today. Since 1999, when the idea was first put forward by the Massachusetts Institute of Technology, many states such as the United States, Japan, China, and the European Union have adopted IoT for various purposes. IoT can efficiently integrate network services into communications and electric power networks and enhance the performance of system usage [15].

3 IOT ARCHITECTURE AND BASIC BLOCKS

The efficacy and utilization of such a system straightforwardly correlates with the quality of its building blocks and how they communicate, as shown in Figure 13.2. There are different approaches to IoT architecture that displays the relationship between different blocks of an IoT system for the collection, storage, and processing of information [16].

- Things: The entity having inbuilt sensors that collect data to be transmitted through a network and actuators that allow objects to function is called Thing. This definition involves houses, cars, fridges, streetlamps, apparatus for manufacturing, and any equipment for rehabilitation that is conceivable.
- Gateways: Information is passed to the cloud through the gateways from objects, and vice versa. It offers synchronization among items; the cloud component of the solution provided by IoT facilitates pre-processing and

Benefits of IoT in Monitoring

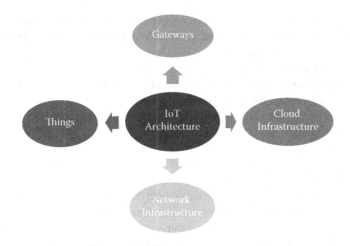

FIGURE 13.2 Basic Blocks of IoT Architecture.

segmentation of data before transferring it to the cloud, and broadcasting control commands from cloud to stuff. Objects use operators to execute commands. The gateway permits data constricted and transfers them between gateways and servers. Information processor streaming ensures proficient transmit of input data to a data lake and control purposes.

- Data Lake/Cloud Infrastructure: It is used to retrieve information in its natural form created by the linked devices. Big data comes in "boxes" or "streams". It is removed from the data lake and inserted into a big data storehouse when the information is required for practical insights.
- Big data warehouse/Network Infrastructure: Processed and non-processed data is stored in a large storehouse and consists of clean, organized, and synchronized data. In addition, the storehouse retrieves background data regarding things and sensors and sends for command control uses.
- Data analytics: It can utilize a big data repository to identify fashion and achieve insights that are actionable. For example, when analyzed, big data display the performance of devices and help make the IoT system more trustworthy and customer-oriented. The relationship and trends manually discovered may also lead to the development of algorithms for control applications.

3.1 Device Management

To confirm a proper IoT device, some procedures are required to control the performance of associated devices facilitating contact among devices, ensuring safe data transmission, and more:

- Device identification to determine the device's identity and ensure it is a genuine device with trustworthy software that transmits reliable information.

- Configuring and monitoring device tuning according to IoT framework purposes. If a computer is mounted, certain parameters must be written down (for example, specific computer ID). Other settings may require changes (e.g., the time between sending messages).
- Testing and diagnostics to confirm the smooth and safe network efficiency of each system and reduce the possibility of breakdowns.
- Updates and repair tools to add functionality, patch bugs, and correct security vulnerabilities.

3.2 User Management

It is essential to have control over users who have access to an IoT system in addition to device management. User management involves defining users—their responsibilities, access rates, and ownership—of a network. This involves options such as adding and deleting users, maintaining user settings, restricting access to certain information by specific users, as well as authorizing certain operations within a program, monitoring and documenting user activities, and more [17].

3.3 Security Monitoring

In the internet of things, security is the chief concern. Connected things generate a huge quantity of data that needs to be transmitted safely, secured from cyber-criminals. Another side is that villains can be entry points for the items linked to the internet. What's more, cyber-criminals can reach and take control of the entire IoT system's "brain." To mitigate these issues, it makes sense to record and analyze commands that are sent to stuff through control applications, track user behavior, and store all these data in the cloud. Using an approach, early-stage security vulnerabilities can be resolved, and steps can be taken to minimize their effect on an IoT system (e.g., ban such commands from control applications). It is also possible to detect trends of suspicious behavior, store these samples, and compare them with the logs created by an IoT system to avoid future intrusion [18].

4 IOT BASED GRID

Energy use and development is an exceedingly difficult job where standard load profiles can help. Since the load curves of the modules are nonlinear, doing the same with microgrids is much more complicated using traditional methods [19]. In microgrids, IoT's ability to produce useful data can significantly increase predictive accuracy. For instance, by supporting streetlights with detectors and connecting them to the network, cities can dim energy-saving light, only restoring them to full power when activity is sensed by the sensors. This would increase infrastructure prices by 70–80% by using various cellular and wired networking protocols such as Zigbee, Wi-Fi, home link, power line carrier, General Packet Radio Service (GPRS), WiMax, LTE, lease cable, and fibers [20]. Grid applications powered by IoT connect in two-way directions. Several software packages are being created for the modern grid operation, maintenance, and management such as

Benefits of IoT in Monitoring

- Customer information systems (CIS)
- Distribution management system (DMS)
- Supervisory control and data acquisition system (SCADA)
- Outage management systems (OMS)
- Geographic information systems (GIS)

4.1 APPLICATIONS IN REAL-TIME SYSTEMS

- Load Forecasting
  Load forecasting involves the extraction of energy consumption trends out of load profile. This can be achieved via classifiers capable of defining strategies within the load curve. To build these classifiers, a microgrid's total load is determined by single units, along with load curves. If use is made of intelligent devices that can monitor their consumption, classification may be more accurate. Therefore, the classification of the smart microgrids in the initial learning process is not needed. Incorporating prediction details to local electricity production planning would result in substantial cost savings, particularly though there are no other smart devices on the entire grid [21]. The incorporation of renewables into the grids is also incredibly significant. For example, the penetration of solar power can be increased if appropriate steps are taken regarding the projection of solar energy.
- Dynamic Pricing
  The next pricing strategy is believed to be competitive pricing, as utility providers will give consumers opportunities to balance the total load [22]. Mobile apps are potentially the building blocks of future mobile microgrids. A smart computer should be capable of the following:

  - Monitor its consumption while achieving its local objectives
  - Interact with other microgrid systems
  - Operates jointly to reach a strategic goal

  A smart meter is an important instrument for dynamic pricing. The metering unit is the interface between the power company's local microgrid and grid.
- Device Collaboration
  A system may be called combined only when the information passed to others is valid and assists to solve the problem of optimization. All intelligent devices load information and pass them to agents who cumulate certain curves, send the highest priority to the users, and inquire to optimize their role in utilization. Smart devices can also work together to help each other achieve the ultimate objective (i.e. reduce energy costs) [23,24]. This would promote innovation and healthy competitiveness because the data generated from the previous phases allow all stakeholders to make informed decisions regarding the usage, production, and potential investments of electricity.
- Smart Meters
  The first and only big step in favor of an intelligent grid allowing IoT is mass deployment of an intelligent indicator—a system attached to a smart

grid's customer side. Smart meters must allow appliances that can work in both positions to be incorporated, such as electric vehicles that can be used to accumulate energy to reduce peak energy requests [25]. A digital smart, bidirectional meter allows two-way communication between utilities and consumers. It registers and records the central utility's electricity usage. Sometimes, the loading operations map a load to a level. The production of energy is modeled on a negative charge. In turn, it provides real-time power consumption data to consumers via a web browser or mobile device, helping them make informed usage choices for the load, achieving demand response management by the utility. Under the Smart Meter National Program (SMNP) of the Ministry of Power (MoP), 1.2 million smart meters were installed in India to date, as shown in Table 13.1 [26].

5 COMPARISON OF CONVENTIONAL POWER GRID AND SMART GRID

Essentially, the modern power system is the association of different elements of the grid-like synchronous devices, transmission substations, power transformers, control lines, transmission lines, control substations, and certain charging forms [27,28]. Such stations are situated far from the load area and energy is delivered using long transmission lines. The smart or intelligent grid is a composite variant of the conventional grid, making electricity grids cleaner and more reliable. It is also a mutual interaction among the energy provider and the energy user. The smart grid can monitor network movements linked to the grid, customer perceptions of energy usage, and deliver real-time information of all events. The key elements of the Electric Grid are smart switches, smart substations, smart meters, and connected synchrophasor networks. Table 13.2 shows the comparison of Conventional Power Grid and Smart Grid.

Effectively deployed and paired with expanded renewable energy efficiency, smart grids will offer many benefits to developed countries especially in terms of reducing power outages and network electrical losses. Table 13.3 lists major

TABLE 13.1
Number of Smart Meters Installed in India

State	No. of Meters Installed
Bihar	26,473
Delhi	57,662
Haryana	1,47,625
Uttar Pradesh	10,09,477
Total	12,41,237

TABLE 13.2
Comparison of Conventional Grid with Smart Grid

Characteristics	Smart Grid	Conventional Power Grid
Generation	Distributed	Centralized
Technology	Digital	Electromechanical
Distribution	Two-way distribution	One-way distribution
Monitoring	Self	Manual
Restoration	Self-healing	Manual
Sensors	Sensors throughout	Few sensors
Control	Pervasive	Limited
Customer Choices	Many	Fewer
Equipment	Adaptive and islanding	Failure and blackout

TABLE 13.3
Benefits of Smart Grid Technology in a Developing Country

Rank	Benefit	Primary Beneficiary
1	Reduced SO_x NO_x and PM10 emissions	Society
2	Reduced CO_2 emissions	Society
3	Reduced sustained outages	Costumers
4	Reduced ancillary service costs	Utility
5	Deferred distribution investments	Utility
6	Reduced equipment failure	Utility

advantages as determined by an IRENA smart grid study in developing nations. The value and culture in general should reap much of the gains. The biggest gain that can be anticipated from the smart grid, though, is a decrease in power outages that would favor energy customers. Decreased outages mean that utility does not lose as much power in extended blackout periods.

6 APPLICATION OF IOT IN THE ELECTRICAL POWER INDUSTRY

In the electrical power sector, IoT assessment changed the way items normally work. For resource usage and costs, IOT expanded the use of wireless technologies to link power sector properties and infrastructure [29–32]. Some examples of IoT use include SCADA, intelligent metering, building automation, smart grid, and wired public lighting.

6.1 IoT SCADA

SCADA is one of IOT's main fields of operation. SCADA enables direct management and control of generation and transmission networks from distant locations. It consists of sensors, actuators, controls, and communication equipment at the remote field site and a central master unit with controlling-side communication systems. This gathers data from field sensors and provides an HMI user interface at the central station.

It also preserves the time-stamped data for further analysis. IoT SCADA is a step above SCADA, which has been used since earlier days. It provides real-time acquisition of the signal and data logging through the internet and IoT server. By understanding its monitoring and control capabilities, it connects human computers, monitors, sensors, and other electrical equipment with the internet.

6.2 Smart Metering

It is a key element in the implementation of smart grids, as they use the Internet of Things to change outdated energy infrastructure. By managing metering operations remotely, smart metering via IoT helps reduce operating costs. It also increases forecasting, decreasing fraud. Those meters simply capture the data and send it back over highly reliable communication infrastructure to utility companies.

6.3 Building Automation

IoT-based software enables property owners to effectively monitor and maintain buildings by linking lighting systems, elevators, environmental controls, and other electrical equipment to the internet and communication technologies. This decreases electricity consumption by shutting off the lights automatically when the rooms are not used and ensuring machines do not waste too much power. IoT-based systems provide remote monitoring and control of smartphones and internet devices for end-users and operators.

6.4 Connected Public Lighting

It is composed of a smart city project that incorporates wireless IoT systems to connect IP-based lighting. This sophisticated street lighting uses intelligently-wired outdoor LED luminaires controlled remotely from the control panel. This design type also allows complicated adjustment of lighting based on changing environmental conditions. This will lead to a significant reduction in operating costs and power consumption.

7 CASE STUDIES

Besides internet access, most new developments in automated and smart engineering and architecture emerge from the Internet of Things (IOT) to become a ground-breaking technology in changing other aspects of everyday life [33].

Although IoT started as a supply chain management, it has grown into a wide range of applications such as transportation, infrastructure, industrial automation, healthcare, construction, and home automation, etc. With IoT, smart communication with current networks without human interference results in omnipresent, computer-sensitive knowledge. Different case studies have been discussed below [34]:

CASE STUDY 1. DUKE ENERGY

Duke Energy, a Florida-based electricity holding firm, claims to have developed a self-healing grid network to instantly reconfigure yourself when you lose control at home. This states the electrical device can identify, isolate, and reroute electricity automatically when a malfunction occurs. Remote smart sensors sense issues at sub-stations and on power lines and connect with the control panel. Instead, switches separate the affected line segment automatically. The control mechanism, in turn, controls the grid condition constantly and decides the right way to transfer electricity to as many citizens as possible. It then reconfigures the electric grid automatically to recover electricity on-site—all in less than a minute. Indiana's state is now modernizing the state-wide electric infrastructure, so expect a verdict on the system in time.

CASE STUDY 2. NATIONAL GRID

The National Grid employs demand-side response service—the Direct Demand system of Free Energi to balance supply and demand through the power grid of the UK. Open Energi says that the network aggregates the use of energy from across customer sites to provide a fast, flexible solution compared to a power station; instead of adjusting supply to match demand, it changes production to meet production. Dynamic Demand is intended to provide the National Grid with a quick response to demand and allow consumers to better monitor their consumption, thus freeing up electricity for the entire grid.

CASE STUDY 3. BROOKLYN MICROGRID

Brooklyn Microgrid has developed an electricity linked network in New York. According to Fast Coexist, the company claims it uses blockchain technologies to allow the first peer-to-peer sharing of resources and a new network internet. The initiative, known as TransActive Grid, is a joint venture between developer LO3 Energy of Brooklyn Microgrid and developer Consen Sys of blockchain technology. The infrastructure includes a smart meter hardware base and a blockchain software base. Participating homes are fitted with smart meters connecting to the blockchain to monitor the energy produced and used in the homes and handle community transactions. As Quick Coexist explains: "Five homes with solar panels produce energy on one side of President Road. On the other hand, five households purchase power from opposite households when they don't require it." A blockchain network is in the center, which handles and tracks transactions with no human contact.

CASE STUDY 4. GENERAL ELECTRIC

General Electric is no newcomer to the IoT Suite. The firm has just picked up $800 million in digital power orders from Asia-Pacific and shows no signs of slowing down its bet on the Internet of Industrial Things (IIoT). For a variety of power plants, General Electric's Resource Quality Management program is used to link various data sources and help interpret the results. Sensors are mounted on main facilities, such as gas turbines, to track and capture operation and output data to help the software tick. The data collected will then be stored in the cloud and provide real-time gas consumption reports inside a facility.

8 FUTURE OF IOT IN ENERGY MARKET

The energy market Internet of Things was valued at USD 15.04 billion in 2019 and is projected to hit USD 39.09 billion by 2025, with a CAGR of 17.24% over the projection period (2020–2025). Researchers expect the Internet of Things (IoT) to greatly impact the energy sector. From tracking a room temperature using sensors to sophisticated systems that regulate an entire building's energy consumption, IoT aims to reduce costs [35].

- The declining natural resources providing electricity and the growing consumption of energy have attracted the interest of various nations around the world to invest in energy waste management strategies and implement high-performance strategies.
- The predictive analytics network tracks the data obtained from sensors and integrated into the power grids. Machinery can, thus, reduce errors in transmission and delivery of electricity to make electricity use worldwide highly efficient and cost-effective.
- Corporations like Nexus Solutions are offering wired technologies to track and reduce energy consumption in buildings. Aquicore is also developing an analytics platform that connects to energy meters already installed. This enables organizations to make decisions to increase the productivity of employees and minimize energy waste.
- Moreover, the incorporation of modern renewable distributed energy systems (DER) into the current grid is intended to deliver locally produced electricity, which is difficult.

9 CONCLUSION

The Internet of Things (IoT) is now an emerging reality in Indian DISCOM operations. Assisted by cloud computing, big data, and AI/ML, the insights derived from data collected from IoT devices can be used in multiple ways to serve customers, enhance efficiency, improve decision-making, prevent breakdowns, and increase asset performance. As more devices connect, Indian DISCOMs need to adopt international best practices in interoperability, data and communications security, and capacity building to enable a seamless transition to mature IoT-enabled systems.

IoT has become a staple of the oil business, making improvements, and delivering undeniable benefits. It offers incentives for the energy provider as well as the energy user the advances in emerging technology and storage, and usage led to the common good of energy conservation. Because of these opportunities, owing to the aging current structures and integration issues IoT also faced certain difficulties in the energy sector. Through recycling processes and cost-cutting practices, IoT will also help produce electricity at an even lower cost. Combined implementation of IoT, AI, and machine learning will revolutionize the entire energy sector by controlling the equipment to produce electricity and minimize jobs.

REFERENCES

1. Ministry of New and Renewable Energy, http://mnre.in/, Accessed on18th March 2020.
2. Central Electricity Authority, http://cea.nic.in/reports.html, Accessed on18th March 2020.
3. Central Electricity Regulatory Commission, http://www.cercind.gov.in/, Accessed on18th March 2020.
4. Li, X., Gong, Q.-W., and Qiao, H. (2010). The application of IOT in power systems. *Power System Protection and Control*, *38*(22), pp. 232–236.
5. Qing-Hai, O., Zheng, W., Yan, Z., Xiang-Zhen, L. and Si, Z. (2013). Status monitoring and early warning system for power distribution network based on IoT technology. In Proceedings of 2013 3rd International Conference on Computer Science and Network Technology (pp. 641–645). IEEE.
6. Mokin, V. B., Skoryna, L. M., Yascholt, A. R. and Kryzhanivskiy, Y. M. (2017). Method for selecting the ranking criteria for monitoring stations of the status of spatially distributed systems and for defining the priority of their location. In 2017 IEEE First Ukraine Conference on Electrical and Computer Engineering (UKRCON) (pp. 870–875). IEEE.
7. Ku, T.-Y., Park, W.-K. and Choi, H. (2017). IoT energy management platform for microgrid. In 2017 IEEE 7th International Conference on Power and Energy Systems (ICPES) (pp. 106–110). IEEE.
8. Al-Ali, A. R. (2016). Internet of things role in the renewable energy resources. *Energy Procedia*, *100*, pp. 34–38.
9. Bekara, C. (2014, August). Security issues and challenges for the IoT-based smart grid. In *FNC/MobiSPC* (pp. 532–537).
10. Srivastava, P., Bajaj, M. and Rana, A. S. (2018, March). IOT based controlling of hybrid energy system using ESP8266. In 2018 IEEMA Engineer Infinite Conference (eTechNxT) (pp. 1–5). IEEE.
11. Wei, L., Miao, W., Jiang, C., Guo, B., Li, W., Han, J., Liu, R. and Zou, J. (2017). Power wireless private network in energy IoT: Challenges, opportunities and solutions. In 2017 International Smart Cities Conference (ISC2) (pp. 1–4). IEEE.
12. Huang, X.-Q., Zhang, J.-Y., Zhu, Y.-S., and Cao, Y.-J. (2013). Research on monitoring system for power transmission and transformation equipments based on IoT. *Power System Protection and Control*, *41*(9), pp. 137–141.
13. Lu, Z.-J., Huang, R., and Zhou, Z.-Y. (2010). Application prospects of internet of things in smart grid. *Telecommunications for Electric Power System* 7.
14. Bedi, G., Venayagamoorthy, G. K. and Singh, R. (2016). Navigating the challenges of Internet of Things (IoT) for power and energy systems. In 2016 Clemson University Power Systems Conference (PSC) (pp. 1–5). IEEE.

15. Shu-wen, W. (2011). Research on the key technologies of IOT applied on smart grid, In 2011 International Conference on Electronics, Communications and Control (ICECC) (pp. 2809–2812). IEEE.
16. Li, X., Huang, Q., and Wu, D. (2017). Distributed large-scale co-simulation for IoT-aided smart grid control. *IEEE Access*, 5, pp. 19951–19960.
17. Sumi, L. and Ranga, V. (2016). Sensor enabled Internet of Things for smart cities. In 2016 Fourth International Conference on Parallel, Distributed and Grid Computing (PDGC) (pp. 295–300). IEEE.
18. Sikder, A. K., Acar, A., Aksu, H., Uluagac, A. S., Akkaya, K. and Conti, M. (2018). IoT-enabled smart lighting systems for smart cities. In 2018 IEEE 8th Annual Computing and Communication Workshop and Conference (CCWC) (pp. 639–645). IEEE.
19. Rhee, S. (2016). Catalyzing the internet of things and smart cities: Global city teams challenge. In 2016 1st International Workshop on Science of Smart City Operations and Platforms Engineering (SCOPE) in partnership with Global City Teams Challenge (GCTC)(SCOPE-GCTC) (pp. 1–4). IEEE.
20. Dalipi, F. and Yayilgan, S. Y. (2016). Security and privacy considerations for iot application on smart grids: Survey and research challenges. In 2016 IEEE 4th International Conference on Future Internet of Things and Cloud Workshops (FiCloudW) (pp. 63–68). IEEE.
21. Al-Turjman, F., and Abujubbeh, M. (2019). IoT-enabled smart grid via SM: An overview. *Future Generation Computer Systems*, 96, pp. 579–590.
22. Lombardi, F., Aniello, L., Angelis, S. De, Margheri, A., and Sassone, V. (2018). A blockchain-based infrastructure for reliable and cost-effective IoT-aided smart grids. 42–46.
23. Gore, R. and Valsan, S. P. (2016). Big Data challenges in smart Grid IoT (WAMS) deployment. In 2016 8th International Conference on Communication Systems and Networks (COMSNETS) (pp. 1–6). IEEE.
24. Shu-wen, W. (2011). Research on the key technologies of IOT applied on smart grid. In 2011 International Conference on Electronics, Communications and Control (ICECC) (pp. 2809–2812). IEEE.
25. Genge, B., Haller, P., Gligor, A., and Beres, A. (2014, May). An approach for cyber security experimentation supporting sensei/IoT for smart grid. In *Second International Symposium on Digital Forensics and Security* (pp. 37–42).
26. https://www.mordorintelligence.com/industry-reports/internet-of-things-in-energy-sector-industry, Accessed on19th June 2020.
27. Bellagente, P., Ferrari, P., Flammini, A. and Rinaldi, S. (2015). Adopting IoT framework for energy management of smart building: A real test-case. In 2015 IEEE 1st International Forum on Research and Technologies for Society and Industry Leveraging a better tomorrow (RTSI) (pp. 138–143). IEEE.
28. Chauvenet, C., Etheve, G., Sedjai, M. and Sharma, M. (2017). G3-PLC based IoT sensor networks for SmartGrid. In 2017 IEEE International Symposium on Power Line Communications and its Applications (ISPLC) (pp. 1–6). IEEE.
29. Kimani, K., Oduol, V., and Langat, K. (2019). Cyber security challenges for IoT-based smart grid networks. *International Journal of Critical Infrastructure Protection*, 25, pp. 36–49.
30. Monnier, O. (2013). A smarter grid with the Internet of Things. *Texas Instruments*, pp. 1–11.
31. Persia, S., Carciofi, C. and Faccioli, M. (2017). NB-IoT and LoRA connectivity analysis for M2M/IoT smart grids applications. In 2017 AEIT International Annual Conference (pp. 1–6). IEEE.
32. Bhoi, S. K., Panda, S. K., Padhi, B. N., Swain, M. K., Hembram, B., Mishra, D., ... Khilar, P. M. (2018, December). Fireds-iot: A fire detection system for smart home

based on iot data analytics. In 2018 International Conference on Information Technology (ICIT) (pp. 161–165). IEEE.
33. Swastika, A. C., Pramudita, R. and Hakimi, R. (2017). IoT-based smart grid system design for smart home. In 2017 3rd International Conference on Wireless and Telematics (ICWT) (pp. 49–53). IEEE.
34. https://internetofbusiness.com/10-examples-showcasing-iot-energy/, Accessed on 19th June 2020.
35. Lazaroiu, C. and Roscia, M. (2017). Smart district through iot and blockchain. In 2017 IEEE 6th International Conference on Renewable Energy Research and Applications (ICRERA) (pp. 454–461). IEEE.

14 Online Clinic Appointment System Using Support Vector Machine

Ch Sanjeev Kumar Dash[1], Ajit Kumar Behera[1], and Sarat Chandra Nayak[2]
[1]Department of Computer Science, Silicon Institute of Technology, Bhubaneswar, India
[2]Department of Computer science and Engineering, CMR College of Engineering & Technology, Hyderabad, India

1 INTRODUCTION

Nowadays, patients in many healthcare [1,2] facilities suffer from long waiting times. Waiting times in healthcare clinics are categorized as "indirect waiting time" and "direct waiting time." Indirect waiting time is mostly expressed in days and is defined as the number of days between appointment request day and appointment day. Direct waiting time is defined as the time that a patient spends in a clinic to see a doctor. Offering appointments with low indirect waiting time is one of the healthcare managers' issues. Long indirect waiting times may bring medical impacts especially for multi-comorbidity and higher priority patients. Long indirect waiting times also increase the no-show probability of patients that decreases the utilization of the healthcare facility. Prioritization of patients based on their comorbidities and characteristics—and deciding which one should get a sooner appointment—is not a simple problem. There are many factors that play important roles in determining the level of urgency of a patient. Machine learning methods provide a decision-making tool in grouping patients into different priority groups more accurately than a human judgment. Considering patients' backgrounds and environments for clustering patients is an important issue that may be ignored by a human. Therefore, having a tool to find a pattern for patients' priority, considering patients' histories and environment factors, helps the healthcare facilities come up with a more accurate priority diagnosis and scheduling.

In this work, we propose a scheduling model where we categorize outpatients into priority groups based on their comorbidities. We used k-mean clustering to prioritize outpatients. Then, we schedule outpatients based on the determined priority classes within a planning horizon. Our study is the first study in literature that combines both machine learning methods to optimize the scheduling process and decrease indirect waiting time of higher priority outpatients in receiving appointments.

The chapter is structured as follows. The SVM and healthcare-related applications are discussed in Section 2. Section 3 presents the different methods used in this work. Section 4 presents the proposed method. Section 5 presents and analyzes the experimental results followed by Section 6 contains concluding remarks.

2 RELATED WORK

Huang et al. [3] developed and evaluated computational models that use electronic health record (EHR) data in predicting the diagnosis and severity of depression, and its response to treatment. Alizadehsani et al. [4] proposed a feature creation method to enrich the dataset. Then, information gain and confidence were used to determine the effectiveness of features on CAD. Typical Chest Pain, Region RWMA2, and age were the most effective ones besides the created features by means of Information Gain. Moreover, Q Wave and ST Elevation had the highest confidence. According to Dash et al. [5], features are ranked according to their importance on clustering, and then a subset of important features is selected. For large data, they used a scalable method using sampling. The aim of El-Bialy et al. [6] is to apply an integration of the results of the machine learning analysis to different data sets targeting the CAD disease. This will avoid the missing, incorrect, and inconsistent data problems that may appear in the data collection. Chaurasia [7] proposed a method to predict more accurately the presence of heart disease with a reduced number of attributes. Originally, 13 attributes were involved in predicting heart disease, and it was reduced to 11 attributes. Three classifiers like Naive Bayes, J48 Decision Tree, and Bagging algorithm are used to predict the diagnosis of patients with the same accuracy as obtained before the reduction of the number of attributes. Dangare et al. [8] designed a model named Heart Disease Prediction system (HDPS). The HDPS system predicts the likelihood of a patient getting heart disease. For prediction, the system uses sex, blood pressure, and cholesterol-like 13 medical parameters. Here, two more parameters are added (i.e. obesity and smoking) for better accuracy. Vijayarani1 et al. [9] used Support Vector Machine (SVM) and Artificial Neural Network (ANN) to predict kidney diseases. The aim of this work is to compare the performance of these two algorithms based on their accuracy and execution time. Lu et al. [10] considered an integrated data mining approach for general sudden deterioration warning. Then, they synthesized a large feature set that includes first- and second-order time-series features, detrended fluctuation analysis (DFA), spectral analysis, approximate entropy, and cross-signal features. After that, they systematically applied and evaluated a series of established data mining methods, including forward feature selection, linear and nonlinear classification algorithms, and exploratory under sampling for class imbalance. Pournajaf et al. [11] proposed a general health deterioration prediction system. They integrated existing mining methods to study several features of two main clinical time-series heart-rate and oxygen saturation. It included features of individual time series such as DFA, entropy, energy, and variance and combined features including correlation and cohesion. Then, they used a forward feature selection method to select the best set of features that contribute to the results. Finally, they applied several classification methods such as SVM and Logistic Regression to train data. Lee et al. [12] developed and proposed a new and unique methodology useful in developing various features of heart rate variability (HRV) and carotid arterial wall thickness helpful in diagnosing cardiovascular disease. We also propose a suitable prediction model to enhance the reliability of medical examinations and treatments for cardiovascular disease. Delen et al. [13] used three popular data mining techniques (decision trees, artificial neural networks, and support vector machines) along with the most commonly used statistical analysis technique—logistic regression—to develop prediction models for prostate cancer survivability. The data set contained around 120,000 records and 77 variables. A k-fold cross-validation methodology was used in model building, evaluation, and comparison. The results showed that support vector machines are the most accurate predictor (with a test set accuracy of 92.85%) for this domain, followed by artificial neural networks and decision trees. Joshi et al. [14] have proposed a method to early diagnose breast cancer patients. V et al. [15] proposed a technique to recognize the symptoms that cause the CKD and pathological blood and urine test indicate the key attributes. They used SVM to classify the dataset. The performance of the proposed scheme is evaluated in terms of sensitivity, specificity, and classification accuracy. Results reveal that an overall classification accuracy of 94.44% was obtained by combining six attributes. It can be concluded that the SVM-based approach is a potential candidate for the classification of chronic kidney disease and non-chronic kidney disease. Marimuthu et al. [16] reviewed heart disease prediction using machine learning and data analytics approach. Begum et al. [17] analyzed and compare the SVM ensemble (ADASVM) with K-Nearest Neighbour (KNN) and SVM classifiers. The leukemia dataset is used as a benchmark to evaluate and compare the performances of ADASVM with KNN and SVM classifiers. Farahmandian et al. [18] diagnosed diabetes using the algorithms of the data that are crucial in diagnosis and prediction. Kausar et al. [19] selected relevant clinical features that can accelerate classification performance to distinguish abnormal and normal patients. For this purpose, the Principal Component Analysis (PCA) algorithm is applied to reduce the attribute dimension by incorporating class identifiers for extracting minimal attributes that have a maximum portion of the total variance. This approach combines Supervised and Unsupervised learning methods namely Support Vector Machines (SVM) and K-means Clustering for classification by adjusting their related parameters and measures. Polat et al. [20] used the Support Vector Machine classification algorithm to diagnose Chronic Kidney Disease. To diagnose Chronic Kidney Disease, two essential types of feature selection methods namely, wrapper and filter approaches, were chosen to reduce the dimension of the Chronic Kidney Disease dataset. In the wrapper approach, classifier subset evaluator with greedy stepwise search engine and wrapper subset evaluator with the Best First search engine were used. In the filter approach, correlation feature selection subset evaluator with greedy stepwise search engine and filtered subset evaluator with the Best First search engine were used. Uzer et al. [21] have proposed a technique to diagnose liver diseases and diabetes commonly observed, using SVM, and develop quality of life. Magerlein and Martin [22] and Cayirli and

Veral [23] provided a comprehensive review of the outpatient scheduling studies. Magerlein and Martin [22] classified the studies into two main groups: "advance scheduling" where patients are scheduled in advance and "allocation scheduling" where available patients are scheduled on the service day. Our scheduling model is an advanced scheduling model. Patrick et al. [24] introduced an advance dynamic scheduling system. The decisions are made at the end of each day, and the outpatients who did not receive an appointment join the next day waiting queue. Chen and Robinson [25] proposed an appointment model of a combination of advance and allocation scheduling. Their models determine when the same-day appointments should be scheduled throughout the day and how these same day appointments affect the routine appointments.

3 BACKGROUND

The background of this research work is presented in this section. We provide an overview of a few concepts our work has been developed and based on. The salient features of feature selection, K-means, and SVM are also discussed distinctly.

3.1 FEATURE SELECTION

Feature selection [5] is an essential task to remove irrelevant and/or redundant features. In other words, feature selection techniques study how to select a subset of potential attributes or variables from a dataset. For a given classification problem, the network may become extremely complex if the number of features used to classify the pattern increases. The reason behind using FS techniques includes reducing dimensionality by removing irrelevant and redundant features, reducing the number of attributes needed for learning, improving algorithms' predictive accuracy, and increasing the constructed model's comprehensibility. After feature selection, a subset of the original features is obtained, which retains sufficient information to discriminate well among classes. The selection of features can be achieved in two ways.

Filter method: It precedes the actual classification process. The filter approach is independent of the learning algorithm, computationally simple, fast, and scalable. Using the filter method, feature selection is done once and can be provided as inputs to different classifiers. In this method, features are ranked according to some criteria, and the top k features are selected.

Wrapper method: This approach uses the method of classification to measure the importance of feature sets; hence, the selected feature depends on the classifier model used. In this method, a minimum subset of features is selected without learning performance deterioration. Different wrapper methods are given below.

  i. Forward feature selection – The procedure starts with an empty set of features (reduced set). The best of the original features is determined and added to the reduced set. At each subsequent iteration, the best of the remaining original attributes is added to the set.

ii. Backward feature elimination – The procedure starts with the full set of attributes. At each step, it removes the worst attribute remaining in the set.
iii. Recursive feature elimination – Recursive feature elimination performs a greedy search to find the best performing feature subset. It iteratively creates models and determines the best or the worst performing feature at each iteration. It constructs the subsequent models with the left features until all the features are explored. It then ranks the features based on the order of their elimination. In the worst case, if a dataset contains N number of features, RFE will do a greedy search for 2N combinations of features.

3.2 K-MEANS CLUSTERING

Partitioning clustering is the most fundamental and simplest method of cluster analysis that arranges objects of a dataset into different exclusive clusters [26]. K-mean clustering is one of the most useful partitional clustering methods. In this method, k points are randomly selected from the data as the center of the clusters. All the other points should be assigned to each cluster using a minimum distance of each point to each centroid. The center of each cluster is updated using the data point within it. All these iterations should be repeated until convergence criteria are met. Euclidean distance is the most useful way to calculate the distance between two points. Let x_i be the centroid of cluster c_i and $d(x_j, x_i)$ be the dissimilarity between point centroid of each cluster and any points belong to that cluster (for all $x_j \in c_i$). Thus, the function to be minimized by k-means can be written as follow:

$$\text{Min } E = \sum_{i=1}^{K} \sum_{x_i c_i^\in} d(x_i, \bar{x}_i) \tag{14.1}$$

3.3 SUPPORT VECTOR MACHINE

Support vector machine (SVM) [27,28] is a supervised machine learning algorithm and is widely used for linear and non-linear classification and non-linear regression. In a nutshell, SVM uses a nonlinear mapping to transform the original training data into a higher dimension. Within this new dimension, it searches for the linear optimal separating hyperplane. With an appropriate non-linear mapping to a sufficiently high dimension, data from two classes can always be separated by a hyperplane (c.f., Figures 14.1 and 14.2).

The SVM finds this hyperplane using support vectors and margins (c.f., Figure 14.2).

SVM is known to be very resistant to the over-fitting problem and achieves a high generalization performance [29,30]. An important property of SVM is that training SVM is equivalent to solving a linearly constrained quadratic programming problem and the solution of SVM is unique and globally optimal, unlike neural networks training that require nonlinear optimization with the danger of achieving local minima [31]. SVM can perform both linear and non-linear classification [32] using different kernel functions that implicitly map their inputs into high-dimensional feature spaces. The original samples may be stated in a finite-dimensional space that

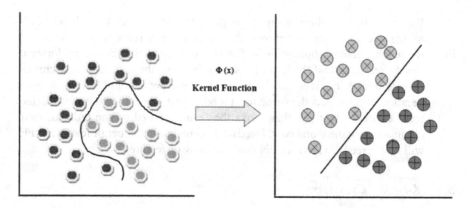

FIGURE 14.1 Mapping by Kernel Function from Non-linearly Separable Space to Linearly Separable Space.

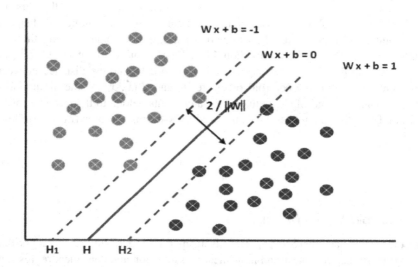

FIGURE 14.2 An Optimal Hyperplane along with Margins and Support Vectors.

is not often linearly separable. Hence, the original finite-dimensional space needs to be mapped into a much higher-dimensional space, making the separation easier. This is done by defining them in terms of a kernel function $K(x_i, x_j)$ selected to suit the problem. The commonly used kernel functions [26] are enumerated in Table 14.1.

Using SVM [33,34] with the above mentioned traditional kernels (Linear, Polynomial, and Gaussian RBF) may not always yield the best classification performance. In such cases, hybridizing or mixing more than one kernel has the potential to improve the classification performance. The hybridization can be achieved by combining two kernels in different proportions. Three such examples are

TABLE 14.1
Different Kernel Functions Used in SVM

Name of the Kernel	Mathematical Formula
Linear kernel	$K(x_i, x_j) = x_i^T \cdot x_j$
Polynomial kernel	$K(x_i, x_j) = x_i^T \cdot x_j^P$ or $K(x_i, x_j) = x_j^T \cdot x_j + 1^P$, where $p > 1$ is the degree of the polynomial
RBF (Gaussian) kernel	$K(x_i, x_j) = \exp - \|x_i - x_j\|^2 / 2\sigma^2$, where σ is the width parameter
Sigmoidal kernel	$\tanh\left(kx_i^T \cdot x_j - \delta\right)$, where k and δ are parameters

$\chi 1$.Linear + $\chi 2$.Polynomial, $\chi 1$.Polynomial + $\chi 2$.RBF, and $\chi 1$.Linear + $\chi 2$.RBF. The mixture of different kernels is beneficial because they may complement each other.

4 PROPOSED METHOD

We divide our proposed method into three phases. In the first phase, wrapper feature selection with k-means clustering is used to divide patients into clusters. In the second, the machine learning technique is used to classify whether the patient with a disease helps give the slots to patients. After classification, a prognosis is given to patients. Finally, patients problem is solved those who not being comfortable to share their disease in front of others. Moreover, for patients who do not understand how to input their symptoms, a feature to video call an intern is provided.

4.1 CLUSTER PATIENTS BASED ON THEIR SYMPTOMS TO PROVIDE SLOT

In our work, k-means algorithm is first done to cluster the patients into high or low priority; the symptoms contain 56 attribute—a very high number to fit in k-means cluster analysis. To reduce the number of features for clustering, we used a feature selection method called wrapper method with a forward selection procedure to prioritize the attributes and rank them according to their relevance. For example, attributes name, age, sex, and registration ID have a lower rank because they hardly contribute to clustering; blood pressure, pulse, DM, HYTN, etc. are important in the clustering mechanism so they rank high. The importance of each feature is calculated using entropy. Dash and Liu [5] proposed the entropy-based ranking for the first time. The entropy for each feature is calculated as follow:

$$E(t) = - \sum_{i=1}^{N} \sum_{j=1}^{N} S_{ij} \log(S_{ij}) + (1 - S_{ij})(1 - S_{ij}) \qquad (14.2)$$

where S_{ij} is the similarity between two points i and j, and it is calculated based on the distance between these two points after feature t is removed ($dist_{ij}$).

$$S_{ij} = e^{-a \times dist_{ij}} \tag{14.3}$$

where, based on [5] α, is assumed to be $\alpha = (\ln(0.5)/dist)$ where dist is the average distance of all points after feature t is removed. After calculation of all features entropy, the best features subset should be determined by calculating the cluster quality. In this study, scattering criteria is used to measure the cluster quality, considering the scatter matrix in multiple discriminant analysis. The within-cluster scatter P_W and between-cluster scatter P_B can be calculated as follows:

$$P_W = \sum_{j=1}^{K} \sum_{x_i c_j^\epsilon} (x_i - m_j)(x_i - x_j)^T$$
$$P_B = \sum_{j=1}^{K} (m_j - m)(m_j - m)^T \tag{14.4}$$

where m is the total mean vector and m_j is the mean vector for cluster j. To evaluate the cluster quality using between-cluster scatter and within-cluster scatter, the "Invariant criterion" is $tr(P-1wP_B)$ used which measures the ratio of between-cluster to within-cluster scatter. If we add an important feature to the subset, $tr(P-1wP_B)$ increases; if we add an unimportant feature to the subset, $tr(P-1wP_B)$ decreases or remains unchanged. Figure 14.3 below shows the procedure of selecting the best feature subset.

After finding the best features subset, k-means clustering is applied on the dataset with selected features to group the patients.

4.2 Gather the Patients' Details and Give a Prognosis

The patients' details, especially their symptoms. are taken into consideration to provide a prognosis. This is done to reduce the diagnosis time, eventually reducing the direct waiting time. Here, we have only considered heart-related diseases which can be further extended to different fields of medical science such as kidney, liver, orthopedics, etc. To classify them into negative and positive classes we used an SVM classifier. Support vector machines, so-called SVM, is a supervised learning algorithm that can be used for classification and regression problems with support vector classification (SVC) and support vector regression (SVR).

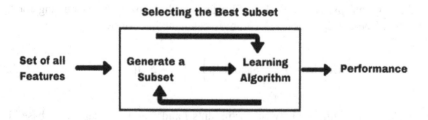

FIGURE 14.3 Wrapper Algorithm for Finding the Best Features Subset.

4.3 VIDEO CALLING FEATURE

The video call to doctor feature is introduced to overcome the problem of patients not comfortable to share their disease in a slot that contains up to 5 patients. Moreover, for patients who do not understand how to input their symptoms, a feature to video call an intern is provided. The video calling feature uses hangout API to connect to the doctor with a provided slot and the intern calling feature is restricted per availability. The intern helps the registered patients provide the symptoms.

Google Hangouts Meet [35] is a standards-based video conferencing application, using proprietary protocols for video, audio, and data transcoding. Google has partnered with Pexip to provide interoperability between the Google protocol and standards-based SIP/H.323 protocols to enable communications between Hangouts Meet and other video conferencing equipment and software.

5 EXPERIMENTAL RESULTS AND ANALYSIS

The experimentations were carried out using the Z-Alizadeh Sani dataset mentioned in this section.

5.1 DATASET DESCRIPTION

The Z-Alizadeh Sani [4] dataset contains the information about the patient we used for our experimental study. A patient is categorized as CAD if his/her diameter narrowing is greater than or equal to 50%; otherwise, it's normal. The dataset contains the records of 303 patients, each with 54 features. The features are arranged in four groups: demographic, symptom and examination, ECG, and laboratory and echo features. Each patient could be in two possible categories: CAD or Normal.

In classifying, the dataset is divided into a training set and a test set. 80% of the datasets were applied for training and 20% were applied for testing. The model and parameter are optimized using the train set. After that, the test dataset is passed to the model (Table 14.2).

5.2 PERFORMANCE METRICS

Several criteria may be used to evaluate the performance of a classification algorithm in supervised learning. A confusion matrix is a useful tool to analyze how well a classifier can identify test samples of different classes, which tabulates the records correctly and incorrectly predicted by the model. Although the confusion matrix provides the information needed to determine how well a classification model performs, summarizing this information with a single number would make it convenient to compare the performance of different models. This can be done using performance metrics such as sensitivity or recall, specificity, precision or positive predictive value, negative predictive value, and accuracy.

Sensitivity: It measures the actual members of the class that are correctly identified as such. It is also referred to as True Positive Rate (TPR) or recall. It is

TABLE 14.2
Dataset Specific Parameters of SVMhk

Data Set	Kernel Parameters for Optimal Accuracy	Kernel Hybridization Parameters	Accuracy Using SVMmk
Z-Alizadeh Sani	= 2	Hybrid of Linear and RBF kernels (0.18 * Linear + 0.82 * RBF)	82.57

defined as the fraction of positive examples predicted correctly by the classification mode.

$$Sensitivity\,(recall) = \frac{TP}{TP + FN} \quad (14.5)$$

Specificity: It is also referred to as a true negative rate. It is defined as the fraction of negative examples that are predicted correctly by the model.

$$Specificity = \frac{TN}{TN + FP} \quad (14.6)$$

Precision (positive predictive value): It is also called positive predictive value and determines the fraction of records that actually turns out to be positive in the group that has been declared as positive by the classifier.

$$Precision = \frac{TP}{TP + FP} \quad (14.7)$$

Negative predictive value (NPV): It is the proportion of the samples that do not belong to the class under consideration and are correctly identified as non-members of the class.

$$NPV = \frac{TN}{TN + FN} \quad (14.8)$$

Accuracy: It is used as a statistical measure of how well a binary classification test identifies or excludes a condition. It is a measure of the proportion of true results.

$$Accuracy = \frac{(TP + TN)}{(TP + FP + TN + FN)} \quad (14.9)$$

where TP = true positives, TN = true negative, FP = false positives, FN = false negatives (Table 14.3).

TABLE 14.3
Outcomes of SVMmk in Percentage

Dataset	Producer Accuracy (Precision) Class-1	Producer Accuracy (Precision)Class-2	User Accuracy (Recall) Class-1	User Accuracy (Recall) Class 2	Overall Accuracy
Z-Alizadeh Sani dataset	70.59	90.23	82.19	82.76	82.57

6 CONCLUSION

In this work, we have used the k-mean clustering method, since it has higher accuracy to group outpatients into two classes of high and low priorities. The wrapper algorithm is used to find the best feature of the dataset used in training the clustering pattern. Therefore, the k-means clustering method can be used to build a pattern and predict the priority class of any incoming outpatient. Whenever an outpatient request arrives, based on the characteristics of the patient, he/she would be assigned to one of the classes. Then, the scheduling model is utilized to schedule the outpatient. The model tries to offer the closest appointments to higher priority outpatients.

REFERENCES

1. Tomar, D. and Agarwal, S. (2013). A survey on data mining approaches for healthcare. *International Journal of Bio-Science and Bio-Technology*, 5(5), pp. 241–266.
2. Almarabeh, H. and Amer, E. (2017). A study of data mining techniques accuracy for healthcare. *International Journal of Computer Applications*, 168(3), pp. 12–17.
3. Huang, S. H., LePendu, P., Iyer, S. V., Tai-Seale, M., Carrell, D., and Shah, N. H. (2014). Toward personalizing treatment for depression: Predicting diagnosis and severity. *Journal of the American Medical Informatics Association*, 21(6), pp. 1069–1075.
4. Alizadehsani, R., Habibi, J., Hosseini, M. J., Mashayekhi, H., Boghrati, R., Ghandeharioun, A., and Sani, Z. A. (2013). A data mining approach for diagnosis of coronary artery disease. *Computer Methods and Programs in biomedicine*, 111(1), pp. 52–61.
5. Dash, M. and Liu, H. (2000, April). *Feature Selection For Clustering. In Pacific-Asia Conference on Knowledge Discovery and Data Mining* (pp. 110–121). Springer, Berlin, Heidelberg.
6. El-Bialy, R., Salamay, M. A., Karam, O. H., and Khalifa, M. E. (2015). Feature analysis of coronary artery heart disease data sets. *Procedia Computer Science*, 65, pp. 459–468.
7. Chaurasia, V. and Pal, S. (2014). Data mining approach to detect heart diseases. *International Journal of Advanced Computer Science and Information Technology (IJACSIT)*, 2, pp. 56–66.
8. Dangare, C. and Apte, S. (2012). A data mining approach for prediction of heart disease using neural networks. *International Journal of Computer Engineering and Technology (IJCET)*, 3(3).

9. Vijayarani, S., Dhayanand, S., and Phil, M. (2015). Kidney disease prediction using SVM and ANN algorithms. *International Journal of Computing and Business Research (IJCBR)*, 6(2).
10. Mao, Y., Chen, W., Chen, Y., Lu, C., Kollef, M., and Bailey, T. (2012, August). An integrated data mining approach to real-time clinical monitoring and deterioration warning. In Proceedings of the 18th ACM SIGKDD International Conference on Knowledge Discovery and Data Mining (pp. 1140–1148).
11. Pournajaf, L. (2013). An integrated data mining approach to real-time clinical monitoring and deterioration warning.
12. Lee, H. G., Noh, K. Y., and Ryu, K. H. (2008, May). A data mining approach for coronary heart disease prediction using HRV features and carotid arterial wall thickness. In 2008 International Conference on BioMedical Engineering and Informatics (Vol. 1, pp. 200–206). IEEE.
13. Delen, D. (2009). Analysis of cancer data: A data mining approach. *Expert Systems*, 26(1), pp. 100–112.
14. Joshi, J., Doshi, R., and Patel, J. (2014). Diagnosis of breast cancer using clustering data mining approach. *International Journal of Computer Applications*, 101(10), pp. 13–17.
15. Ravindra, B. V., Sriraam, N., and Geetha, M. (2018). Classification of non-chronic and chronic kidney disease using SVM neural networks. *International Journal of Engineering & Technology*, 7(1.3), pp. 191–194.
16. Marimuthu, M., Abinaya, M., Hariesh, K. S., Madhankumar, K., and Pavithra, V. (2018). A review on heart disease prediction using machine learning and data analytics approach. *International Journal of Computer Applications*, 975, p. 8887.
17. Begum, S., Chakraborty, D., and Sarkar, R. (2015, December). Cancer classification from gene expression based microarray data using SVM ensemble. In 2015 International Conference on Condition Assessment Techniques in Electrical Systems (CATCON) (pp. 13–16). IEEE.
18. Farahmandian, M., Lotfi, Y., and Maleki, I. (2015). Data mining algorithms application in diabetes diseases diagnosis: A case study. *MAGNT Research, Technical Report 3*(1), pp. 989–997.
19. Kausar, N., Abdullah, A., Samir, B. B., Palaniappan, S., AlGhamdi, B. S., and Dey, N. (2016). Ensemble clustering algorithm with supervised classification of clinical data for early diagnosis of coronary artery disease. *Journal of Medical Imaging and Health Informatics*, 6(1), pp. 78–87.
20. Polat, H., Mehr, H. D., and Cetin, A. (2017). Diagnosis of chronic kidney disease based on support vector machine by feature selection methods. *Journal of medical systems*, 41(4), p. 55.
21. Uzer, M. S., Yilmaz, N., and Inan, O. (2013). Feature selection method based on artificial bee colony algorithm and support vector machines for medical datasets classification. *The Scientific World Journal*, 2013, Article ID 419187, 10 pages.
22. Magerlein, J. M. and Martin, J. B. (1978). Surgical demand scheduling: A review. *Health services research*, 13(4), p. 418.
23. Cayirli, T. and Veral, E. (2003). Outpatient scheduling in health care: A review of literature. *Production and Operations Management*, 12(4), pp. 519–549.
24. Patrick, J., Puterman, M. L., and Queyranne, M. (2008). Dynamic multipriority patient scheduling for a diagnostic resource. *Operations Research*, 56(6), pp. 1507–1525.
25. Chen, R. R. and Robinson, L. W. (2014). Sequencing and scheduling appointments with potential call-in patients. *Production and Operations Management*, 23(9), pp. 1522–1538.

26. Han, J., Pei, J., and Kamber, M. (2011). *Data mining: Concepts and techniques*. Elsevier.
27. Scholkopf, B., Sung, K. K., Burges, C. J. C., Girosi, F., Niyogi, P., Poggio, T., and Vapnik, V. (1997). Comparing support vector machines with Gaussian kernels to radial basis function classifiers. *IEEE Transactions on Signal Processing, 45* (11), pp. 2758–2765.
28. Lin, S., Liu, Z. (2012). Parameter selection in SVM with RBF kernel function, *Journal-Zhejiang University of Technology, 35*(2), p. 135.
29. Wu, C. H., Tzeng, G. H., Goo, Y. J. and Fang, W. C. (2007). A real-valued genetic algorithm to optimize the parameters of support vector machine for predicting bankruptcy. *Expert Systems with Applications, 32*(2), pp. 397–408.
30. Huang, C. L. and Wang, C. J. (2006). GA-based feature selection and parameters optimization for support vector machines. *Expert Systems with applications, 31*(2), pp. 231–240.
31. Belousov, A. I. and Verzakov, S. A. and Von Frese, J. (2002). A flexible classification approach with optimal generalization performance: Support vector machines. *Chemometrics and Intelligent Laboratory Systems, 64*(1), pp. 15–25.
32. Mo, Y. and Xu, S. (2010, June). Application of SVM based on hybrid kernel function in heart disease diagnoses. In 2010 International Conference Intelligent Computing and Cognitive Informatics (ICICCI) (pp. 462–465). IEEE.
33. Dash, C. S. K., Sahoo, P., Dehuri, S., and Cho, S. B. (2015). An empirical analysis of evolved radial basis function networks and support vector machines with mixture of kernels. *International Journal on Artificial Intelligence Tools, 24*(4), p. 1550013.
34. Sahoo, P., Behera, A. K., Pandia, M. K., Dash, C. S. K., and Dehuri, S. (2013, September). On the study of GRBF and polynomial kernel based support vector machine in web logs. In 2013 1st International Conference on Emerging Trends and Applications in Computer Science (pp. 1–5). IEEE.
35. Mollett, A. (2013). Using Google hangouts for higher education blogs and workshops. *Impact of Social Sciences Blog*.

15 Electric Vehicle

Developments, Innovations, and Challenges

Sahil Mishra[1], Sanjaya Kumar Panda[2], and Bhabani Kumari Choudhury[3]

[1]Department of Computer Science and Engineering, Indian Institute of Information Technology, Design and Manufacturing, Kurnool, India
[2]Department of Computer Science and Engineering, National Institute of Technology, Warangal, India
[3]Department of Electrical Engineering, Indian Institute of Technology, Roorkee, India

1 INTRODUCTION

An electric vehicle (EV) uses electric or traction motors in place of internal combustion engines for propulsion [1–3]. It gets power from rechargeable batteries, which are placed inside the vehicle itself [4]. There are several ways to recharge the batteries; most people usually use the household power supply to recharge them. Charging equipment is usually installed in the garage or outside. In a few countries, there are charging stations where vehicle batteries can be charged within 30 minutes. The general cost of a full charge varies from $2.64 to $3 [5]. Other than batteries, vehicles may be self-contained with solar panels or electric generators, which may convert fuel to electricity and charge the batteries. All means of transport—roadways and railways—use EVs like electric cars, electric scooters, and electric trains. While cars and scooters have batteries installed in them, the trains draw power from the overhead lines along with an on-board power supply like a battery or supercapacitor [6]. However, airways and waterways are still lagging a lot behind in terms of EVs due to the huge demand for power supply and the unavailability of constant power supply sources.

2 TYPES OF EVS

The extent of electricity used by EVs is one of the bases of their categorization, which divides them into three broad categories: battery EV (BEV), plug-in hybrid EV (PHEV), and hybrid EV (HEV).

3   BATTERY EV

Battery EV is completely driven by rechargeable batteries, without any gasoline-powered internal combustion engine. It stores the electric power in high-capacity batteries that run the motors and all other components [7,8]. The BEV is usually charged by some external energy source. Various BEVs include Chevy Bolt, Chevy Spark, Nissan LEAF, Ford Focus Electric, Tesla X, and Volkswagen e-Golf.

4   PLUG-IN HYBRID EV

Plug-in Hybrid EV is powered by batteries and internal combustion engines. The batteries are recharged either by plugging into an external power source and by their internal combustion engines. PHEV models can easily run for 10 to 40 miles before their fuel engines start running [7]. Some PHEV are Chevy Volt, Chrysler Pacifica, Ford Fusion Energi, Audi A3 E-Tron, and BMW i8.

5   HYBRID EV

Hybrid EV is also powered by both batteries and internal combustion engines. However, the batteries are recharged through regenerative braking; there is no plug-in way to charge the batteries. HEV starts using the batteries, but the internal combustion engine starts to work when there is a load on the vehicle or its speed increases [7].

6   HISTORY AND RECENT DEVELOPMENTS

Sibrandus Stratingh made a small electric car in the Netherlands in 1835 [9]. Afterward, many advancements had been made in the nineteenth century itself. After two years of the invention of the electric car by Sibrandus, Robert Davidson in Scotland invented the electric locomotive. At the same time, Moritz Von Jacobi invented a basic electric boat in Russia, which could travel at the top speed of 3 miles per hour [10]. During the same period, similar patents were filed by Colten and Lilley in 1847 for the use of track rails as electric current conductors [11]. The first practical electric cars came into existence only after 1859, when the lead-acid battery was invented by French physicist Gaston Planté; the battery capacity was not so good. Therefore, the design of the battery was improved by Camille Alphonse Faure in 1881, which was later manufactured on an industrial scale [12]. In the middle of the twentieth century, the use of electric cars declined due to their short-range capacity. During that time, Konstantin Tsiolkovsky introduced the concept of electric propulsion. Later, in the twenty-first century, the developments and enhancements in EVs started again after Tesla Motors started working on Tesla Roadster in 2004. This car was the first electric car to travel more than 200 miles on one full charge. Tesla released another vehicle, Model S, in 2012. Large automobile companies like Nissan and General Motors stated that it was Roadster which initiated the spark in the so-called dead EV industry [13]. Consumer reports declared Tesla as the best brand in the USA and eighth-best in the world [14]. Electric scooters came before practical electric cars. Initially, the electric bikes came, which were considered as upgraded bicycles. Charger electric bicycle invented the first electric bike with a pedelec in 1997 [15]. But later, they became the replacements for motor-driven scooters. China is the leading manufacturer of e-bikes globally, selling 22.2 million units in 2009 [15] and 9.4 million of the total 12 million in 2013 [16].

7 ADVANTAGES OF EVS

EVs were thought to be made for a single purpose: the protection of the environment. Gasoline-powered internal combustion engines emit massive amounts of pollutants in the atmosphere, causing air pollution and global warming. It is not only air pollution; when it rains, pollutants are dissolved in water and reach water bodies and soil, polluting them as well [17,18]. The BEVs, on the other hand, do not emit such pollutants in the atmosphere, while hybrid EVs optimize the usage of both battery-powered engine and gasoline engine when traveling for longer distances to reduce the emission of such pollutants. The batteries can be easily recycled, reducing carbon footprint [17]. EVs do not even require a lot of maintenance compared to gasoline-powered vehicles. The latter's engines and other parts require lubrication over a while to work smoothly, while EVs require cheaper and less frequent maintenance [17].

Fuel prices are very unpredictable and fluctuate regularly, which makes driving a gasoline-powered car an expensive asset. Electricity prices are stable, and it is much cheaper than conventional fuels. Also, electricity can be easily produced in homes using solar energy, which can even be used to charge the vehicles, cutting off monthly electric charges and fuel costs completely [17]. Nowadays the government has started offering incentives like tax exemptions to people who invest in electric cars [18].

8 DISADVANTAGES AND CHALLENGES FACED BY EVS

The EV industry is still in the development phase. So, a lot of places do not have recharge stations. It remains a big problem for people traveling long distances. Also, the recharge time of batteries varies, generally from 3-8 hours. So, even if there are strategically located recharge stations, the user will have to wait for an exceedingly long time to get the batteries recharged [19]. The efficiency of the batteries is also a big problem due to the weight of the EVs. The batteries of an EV may weigh around 450 kilograms. This weight results in significant pressure on the batteries, leading to faster draining [19]. Batteries are filled with electrolytes that would cause damage to the car and fire when leaked [20].

Along with these problems, the main concern remains, which is the distance traveled. EVs can run for around 100 miles on one full charge. To solve this problem, hybrid cars have been developed that can switch between electric motors and internal combustion engines. But this does not solve the issue, completely as the fuel to run the internal combustion engine again causes pollution [21].

To get rid of most of the above problems, many changes have been introduced in the vehicles. Lithium-ion batteries have improved a lot to store and provide power for a longer period, but this increases the cost of the batteries and vehicles too. These batteries also need timely replacement every 4-5 years, which is again a costly affair [21]. EVs also prevent noise pollution to a considerable extent but this lesser noise emission has a big disadvantage as people around the vehicles would not be able to hear the vehicle coming towards them, which leads to fatal accidents most of the time [22].

Another big challenge faced by EVs is the stereotyped thinking of people that manual transmission vehicles are superior to vehicles with automatic transmission. Most EVs do not have a manual transmission gearbox, so people tend to not buy them. In many regions of the world, heavy vehicles like sports utility vehicles (SUVs) are considered superior to light motor vehicles because of the sound produced by their engines. Therefore, people refrain from buying lighter and less noise-producing vehicles like EVs [18].

9 INNOVATIONS REQUIRED IN EVS

These challenges have been faced by EVs for a very long time and researchers are working on how to get rid of most of the hurdles in the way of EVs. Charging the EVs remain the biggest problem. More specifically, a shortage of electricity is one of the factors. So, governments should go for non-conventional sources of energy to produce electricity like wind energy, solar energy, and tidal energy. This could reduce the problem of shortage of electricity to a certain extent [23]. Not only that, but the cost of electricity could also be cut by a great margin. Recharge stations need to be installed in multiple places around the globe. Solar cars have not been a great success due to the limited charging capacity of solar energy. If the solar panels are embedded into the whole body of the car, this would increase the surface area and may provide better charging capacity. EV manufacturers can install their private recharging stations and charge the users per use [24]. Electrolyte leakage can easily be avoided by replacing Lithium-ion batteries with solid-state batteries that are made of solid components. They provide extended lifetime, large temperature range with minimal requirement of cooling mechanism [25].

Wind energy can also be used to charge running EVs. The usual speed at which an EV move is somewhat greater than 50 km/hr. At this speed, the air drag is worth considering. If a turbine can be fixed into the EV, that could charge the moving EV to some extent. If all the above features can be combinedly introduced in the EVs, then a lot of optimization can be brought with them.

Another charging concept can be developed by taking a new smartphone charging system into consideration, which uses induction produced by electromagnetic waves to transfer energy from the charger to the smartphone [20]. These charging systems can be implemented in the surface of the parking. Therefore, when the car is parked over that system, it can be charged. The problem with this system is that it charges smartphones relatively slower than wired chargers; the same would be the problem with cars too. The problems caused by lesser noise emission from EVs can be solved by introducing a system that automatically produces some sort of noise on seeing pedestrians around it. This technology is being used in self-driving cars to detect pedestrians around the vehicles. It helps the cars avoid accidents.

10 CONCLUSION

In this chapter, we have discussed EVs—their broad categorization, and technologies used by them. It also states the future innovations required in EVs needed to solve multiple environmental issues and for energy conservation. There are a lot of hurdles in their way to success. People need to change their behavior towards this disruptive technology. Breaking out of the old and conventional norms is needed to protect the earth. This technology certainly has more benefits than drawbacks and is of utmost need for the earth.

REFERENCES

1. Zheng, J., Sun, X., Jia, L. and Zhou, Y. (2020). Electric passenger vehicles sales and carbon dioxide emission reduction potential in China's leading markets, *Journal of Cleaner Production*, Elsevier, *243*, pp. 1–20.
2. Chakraborty, D., Bunch, D., Lee, J. and Tal, G. (2019). Demand drivers for changing infrastructure-charging behavior of plug-in electric vehicle commuters, *Transportation Research Part D: Transport and Environment*, Elsevier, *76*, pp. 255–272.
3. Naumanen, M., Uusitalo, T., Huttunen-Saarivirta, E. and Have, R. (2019). Development strategies for heavy duty electric battery vehicles: Comparison between China, EU, Japan and USA, *Resources, Conservation and Recycling*, Elsevier, *151*.
4. Electric Vehicle, https://en.wikipedia.org/wiki/Electric_vehicle, Accessed on 8th October 2019.
5. Alternative Fuels Data Center, https://afdc.energy.gov/fuels/electricity_charging_home, Accessed on 4th October 2019.
6. Electric Locomotive, https://en.wikipedia.org/wiki/Electric_locomotive, Accessed on 9th October 2019.
7. Types of Electric Vehicle, https://www.evgo.com/why-evs/types-of-electric-vehicles/, Accessed on 7th October 2019.
8. How Do Battery Electric Cars Work?, https://www.ucsusa.org/clean-vehicles/electric-vehicles/how-do-battery-electric-cars-work, Accessed on 10th October 2019.
9. Sibrandus Stratingh, https://www.rug.nl/university-museum/history/prominent-professors/sibrandus-stratingh, Accessed on 9th October 2019.
10. Electric Boat, https://en.wikipedia.org/wiki/Electric_boat, Accessed on10th October 2019.
11. Electric Traction, A Vital Development in the History of the Railway, http://mikes.railhistory.railfan.net/r066.html, Accessed on 10th October 2019.
12. History of the Electric Vehicle, https://en.wikipedia.org/wiki/History_of_the_electric_vehicle, Accessed on 11th October 2019.
13. Plugged In, https://www.newyorker.com/magazine/2009/08/24/plugged-in, Accessed on 10th October 2019.
14. Which Car Brands Make the Best Vehicles, https://www.consumerreports.org/cars-driving/which-car-brands-make-the-best-vehicles/, Accessed on 12th October 2019.
15. History of the Electric Vehicle, https://en.wikipedia.org/wiki/History_of_the_electric_vehicle, Accessed on 10th September 2019.
16. Electric Motorcycles and Scooters, https://en.wikipedia.org/wiki/Electric_motorcycles_and_scooters, Accessed on 4th October 2019.
17. Advantages of Electric Cars – Top Benefits of EVs https://www.energysage.com/electric-vehicles/advantages-of-evs/, Accessed on 11th October 2019.
18. Advantages and Disadvantages of Electric Cars, https://thenextgalaxy.com/the-advantages-and-disadvantages-of-electric-cars/, Accessed on 11th October 2019.

19. Top 7 Disadvantages of Electric Cars, https://autowise.com/top-7-disadvantages-of-electric-cars/, Accessed on 8th October 2019.
20. Inductive Charging, https://en.wikipedia.org/wiki/Inductive_charging#Disadvantages, Accessed on 12th September 2019.
21. Advantages and Disadvantages of Electric Cars, https://www.conserve-energy-future.com/advantages-and-disadvantages-of-electric-cars.php, Accessed on 12th October 2019.
22. Electric Cars Gears, https://www.greenoptimistic.com/electric-cars-gears/, Accessed on 5th October 2019.
23. Non-Conventional Sources of Energy, https://www.toppr.com/guides/physics/sources-of-energy/non-conventional-sources-of-energy/, Accessed on 12th October 2019.
24. Utilities Electric Vehicle Cost Efficient, https://www.scientificamerican.com/article/utilities-electric-vehicle-cost-efficient/, Accessed on 13th October 2019.
25. Five Emerging Battery Technologies for Electric Vehicles, https://www.brookings.edu/blog/techtank/2015/09/15/five-emerging-battery-technologies-for-electric-vehicles/, Accessed on 24th September 2019.

16 Usage of Convolutional Neural Networks in Real-Time Facial Emotion Detection

Ch. Sanjeev Kumar Dash[1], Ajit Kumar Behera[1], Sarat Chandra Nayak[2], and Satchidananda Dehuri[3]

[1]Department of Computer Science, Silicon Institute of Technology, Bhubaneswar, India
[2]Department of Computer science and Engineering, CMR College of Engineering & Technology, Hyderabad, India
[3]Department of Information and Communication Technology, Fakir Mohan University, Vyasa Vihar, Balasore, Odisha, India

1 INTRODUCTION

Being humans, we often express our emotions for social communication through a variety of means such as body language, voice inflection, EEG, etc. Nevertheless, the simplest and most practical approach is the study of facial expression. The most universally recognized category of emotions is anger, disgust, fear, happiness, sadness. Interestingly, a more complex blend of emotions could be used as descriptors to create other shades. Therefore, the detection of emotion effectively from facial impressions would be considered significant. Such development is invariably used in different areas such as medicine, marketing, and entertainment.

In the current scenario, CNN is the most popular deep learning architecture for image classification and reorganization when compared to the conventional neural network. It achieves a reduction in several parameters by introducing the concept of convolution and pooling through downsampling. CNN can detect key features of an image from the input data automatically without human intervention. It performs parameter sharing using special convolution and pooling operations that make it computationally efficient and able to run on any device. The two major components of CNN are feature extraction followed by classification. The former can be achieved by convolution and pooling layers while the latter is fulfilled by fully connected layers. This is in contrast with the conventional neural net where feature extraction needs human supervision. CNN can capture the spatial and temporal dependencies in an image through appropriate filters. The multilayer perceptron (MLP) is the most commonly used neural architecture. However, it has certain limitations in considering image processing. It uses one perceptron per input pixel. For larger images, the number of trainable weights becomes unmanageable, which may result in overfitting. Therefore, MLPs are not the best approach for image processing.

Usually, emotion identification observes stationary images of facial appearance. Instead, CNN-based emotion recognition in real-time with video stream is investigated in our proposed method. In the proposed approach for detecting human emotion in real-time, different scenes, angles, and lighting conditions are considered. The proposed result is superimposed over other techniques.

The chapter is structured as follows. CNN related applications are talked about in Section 2. The experimental data are presented in Section 3. HAA wavelet is discussed in Section 4. Section 5 presents the CNN method used in this work. Section 6 presents and analyzes the experimental results followed by concluding remarks.

2 RELATED WORK

In recent years, many advanced computer vision techniques have been developed to recognize the facial emotion of humans. Some of the approaches include pyramid histograms of gradients (PHOG), AU aware facial features, boosted LBP descriptors, and RNNs [1–4]. G. Levi et al. used a static image with CNN and found significant improvement in facial emotion recognition [5–8]. The authors used the Local Binary Patterns (LBP) transformation method with small training deep CNN for proper illumination of the image. The intermediate results are applied on a large CASIA Web Face dataset for facial emotions released for the EmotiW 2015 challenge. The results showed better accuracy over baseline models. Lee et al. [9] used three benchmark datasets for classification using CNN. For the classification of the soundtrack of a dataset, Hershey et al. took 70M training videos for video image classification [10]. Differing the training set and vocabulary label, they found the CNN-based model performed better in image classification as well as audio classification tasks. Wang et al. suggested CNN-RNN for image classification and demonstrated that the proposed approach achieves better performance than traditional models [11]. Han et al. has suggested a novel two-phase CNN-based web data augmentation model which could efficiently be used in the original dataset that can be augmented with the most valuable internet images for classification [12]. Their method noticeably expands the training dataset. Li et al. [13] used customized CNN with a shallow convolution layer to classify lung image patches with interstitial lung disease (ILD). The successful applications of CNN in different areas, along with their accuracy in percentage, have been summarized in Table 16.1. Alongside, Table 16.2 summarizes the applications of CNN in different subareas of face recognition.

3 DATASET PREPARATION

We have created a dataset of 700 images categorized by the four mentioned emotions (i.e. sad, happy, calm, and angry). The image resolution was 300 × 240. The groups are Happy, Sad, Angry, and Calm. The dataset is the mix of images of all these four groups; their images were downloaded from the internet in bulk of nearly 150-160 images at a

TABLE 16.1
State-of-the-Art Research Works on CNN Application

Sl.No.	Authors	Year	Dataset	Accuracy
1.	Spanhol et al. [14]	2015	Dataset of histopathological images	85.3%
2.	Wei et al. [15]	2014	Visual Object Classes Challenge (VOC) datasets	93.2%
3.	Knag et al. [16]	2014	Tobacco dataset	65.35%
4.	Xie et al. [17]	2015	Stanford Cars dataset	42.3%
5.	Romero et al. [18]	2015	UC Merced dataset	84.53%
6.	Paoletti et al. [19]	2018	Indian Pines dataset:	99.13%
7.	Ioannou et al. [20]	2016	ImagageNet Large Scale Visual Recognition Challenge object classification dataset	91.88%
8.	Huang et al. [21]	2015	Indian Pines Salinas	90.16% 92.60%
9.	Affonso et al. [22]	2017	Wood samples. Source: Own Authors	79.07%
10.	Barat et al. [23]	2016	15 scenes	91.57
11.	Li et al. [24,25]	2020	3D pavement images	94%

time all belonging to the same group. This was done for all the groups and later, all the images were kept at a single place to make the dataset. A bulk downloader extension was used in chrome to download the images. The bulk downloader downloads all the images present on a webpage. The images were crawled using a bulk image downloader. These images are used for further training of the test case. The images are then trained and classified according to their respective emotions using the HAAR cascade and Convolutional Neural Network [47–52]. The images are then tested in two ways: finding the human face using HAAR Cascade and finding the emotion of the detected human face(s) using CNN.

4 MATERIALS AND METHODS

This segment is divided into two parts. The first describes the HAAR cascade and the second discusses the general architecture of the CNN model.

4.1 HAAR Cascade or HAAR Wavelet

HAAR cascade is an object detection technique used to recognize objects in an image or video. The algorithm has four stages:

a. Haar feature selection
b. Creating integral images

TABLE 16.2
State-of-the-Art Research Works Based on CNN for Face Detection

Sl.No.	Author	Model	Accuracy	Dataset
1.	Arriaga et al. [26]	CNN and Back-propagation	96%	Internet Movies Database gender dataset.
2.	Matsugu et al. [27]	Rule-based CNN	97.6%	Facial fragment images used is 2900
3.	Jiang et al. [28]	R-CNN (Region-based CNN)	95.2%	WIDER face dataset (Web image dataset for event recognition)
5.	Qin et al. [29]	Region proposal network	98.7%91.2%	AFW database
6.	Zhu et al. [30]	(CMS-RCNN)	90.2%	WIDER FACE Dataset
7.	Sun et al. [31]	Region-CNN	80%	WIDER FACE, FDDB
8.	Hao et al. [32]	CNN	99%	Annotated Facial Landmarks in the Wild (AFLW) dataset
9.	Farfade et al. [33]	R-CNN	95%	PASCAL VOC
10.	Li et al. [34]	C-CNN	98%	FDDB,
11	Wang et al. [35]	R-CNN	98.74%	FDDB and WIDER FACE
12.	Zhang [36]	Two-stream contextual CNN architecture	95.92%	WIDER FACE dataset
13.	Ranjan et al. [37]	CNN	97.6%	FDDB datasets.
14.	Szlávik et al. [38]	CNN	Nose-97%, Eyes- 89% Mouth- 92%	University of Manchester Institute of Science and Technology
15.	Mori et al. [39]	CNN using SVM	100%	Database (e.g., Softpia Japan)
16.	Raghavendra et al. [40]	D-CNN		Morphed face image database.
17.	Zhang et al. [41]	R-CNN	95.9%	WIDER FACE dataset
18.	Yang et al. [42]	Deep-CNN	90.99%	CelebFaces dataset
19.	Wan et al. [43]	R-CNN		FDDB dataset
20.	Bansal et al. [44]	CNN		Chinese Academy of Sciences Institute of Automation data
21.	Balya et al. [45]	Analogic cellular neural network algorithms.		The Facial Recognition Technology (FERET) database
22.	Zhang et al. [46]	Feature Agglomeration Networks	98.7%	WIDER

(Continued)

TABLE 16.2 (*Continued*)

Sl.No.	Author	Model	Accuracy	Dataset
23.	Li et al. [24]	CNN	97.38% and 97.18%	Cohn–Kanade (CK+) database and JAFFE database

    c. Adaboost training
    d. Cascading classifiers

From the positive and negative images of the faces, the process of training is started for classification. Then, features are extracted from it. These features are a sequence of rescaled square-shaped features that are like convolutional kernels. There are two kinds of features present here: *edge* feature and *line* feature. The edge feature is used to effectively detect edges such as nose and mouth; line features are used to detect parts like eyebrows, lips, etc. Ideal HAAR feature has a value of 1 assigned to black and 0 to light areas whereas, in a real scenario, it has somewhere from 0.1-0.9 according to darkness measure.

In the second step image, the integration process is used to speed up feature selection. Whatever the features selected, in this step, most of them are irrelevant.

In the third step, Adaboost is used to select the best features and train the classifiers that use them. This process computes a strong classifier by linear combining weighted sum weak classifiers.

During the detection phase, the input image is generated by sliding a window over the target size. Each window of the image Haar features are computed using Viola-Jones face detection given in Equation (16.1).

$$\Delta = dark - white = \frac{1}{n} \sum_{dark}^{n} 1(x) - \frac{1}{n} \sum_{white}^{n} 1(x) \qquad (16.1)$$

were Δ for ideal HAAR feature is 1 and Δ for real scenario is less than 1. This difference is then compared to a learned threshold that separates non-objects from objects.

Each stage in the cascade classifier is an ensemble weak learner called decision stumps. Each stage is trained using a technique called boosting. Boosting techniques train a highly accurate classifier by taking a weighted average of the decisions made by the weak learners. Each stage of the classifier labels the region either positive or negative. *A positive* label indicates the presence, whereas a *negative* indicates the absence of objects. If the label is negative, the classification of this region is complete and the window slides one step towards right to the next location. If the label is positive, the classifier passes the region to the next stage. In this case, the detector reports an object found at the current window location. The negative samples are rejected as soon as possible. The basic hypothesis is that the faraway objects are less informative and do not contain the object of attention. On the other hand, true positives are unusual and worth taking; a false *positive* occurs when a negative sample is wrongly classified as positive at the time of verification.

To work smoothly, each stage in the cascade must have a low false-negative rate. If a stage wrongly labels an object as negative, the process stops and cannot be corrected. However, each stage can have a high false-positive rate. Even if the detector wrongly labels a non-object as positive, the correction can be done in the subsequent stage. By increasing more stages, the overall false-positive rate can be reduced but it also reduces the overall true positive rate.

Finally, by using the Viola-Jones algorithm, we will compare how close the real scenario is to the ideal case.

4.2 Convolutional Neural Network

CNN [21] is composed of a set of blocks that is used across space and time. Each block converts the input pattern to an output pattern of activation functions that will serve as input to the next block. In contrast to conventional ANNs, the blocks of neurons in CNN's operate like kernels that are connected and applied over one patch of the input volume—that is, the neurons of a block are loosely connected to neurons of the previous layer as in standard MLP. A typical CNN with four convolution layer, two fully connected layer, and one output layer is shown in Figures 16.1 and 16.3. The block is comprised of a feature extraction stage, which consists of three layers described as (Figures 16.2–16.4):

A. **Convolution layer:** A 3D layer where each neuron computes the dot product between its weight and a small region of the input volume. The layers as a set of k filters of size $l \times l \times q$ where the neurons share the same weights and bias and connect the input volume to the output volume. Each filter detects a feature at every location on the input layer. The resulting output volume of the first layer is the feature map of size $d^l \times d^l \times k^l$ that stores the information where the feature occurs in the original input volume and is calculated as

$$z_i^l = B^l + \sum_{j=1}^{k^{l-1}} W_{i,j}^l * Z_j^{l-1}, \qquad (16.2)$$

where $i \in [1, k^l]$, B^l is the bias matrix of layer 1 and $W_{i,j}^l$ is the weight matrix of filter that connects the jth feature map in layer l-1 ..with ith feature map in layer l.

B. **Nonlinearity layer:** This layer embeds a nonlinear function that is applied to each feature map's component to learn nonlinear representations: $a^l = f(z^l)$.

C. **Pooling layer:** This layer is used to make the features invariant from the location and to summarize the output of multiple neurons in convolution layers through a polling function. In our case, this layer executes a max operation within a small spatial region R over the resulting feature map after the nonlinearity layer:

$$p^l = \max_{i \in R} a_i^l \qquad (16.3)$$

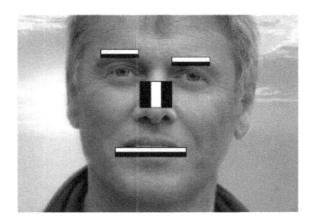

FIGURE 16.1 HAAR Classified Human Face.

5 EXPERIMENTAL RESULTS AND ANALYSIS

The experiments were carried out using the dataset mentioned in Section 4. The webcam took a snapshot of the face and then the image was processed by the same algorithms that were used for training. After processing, the image was classified into one of the emotion classes. We have used Tensorflow for our work. It was imported in Python to use CNN methods. For the classification [53] task, the dataset is divided into two sets—training and test set. 80% of the datasets were applied for training and 20% were applied for testing. The model and parameter are optimized using the trainset. After that, the test dataset is passed to the model. The performances of the model were measured through four statistics such as accuracy, confusion matrix, F1 score, and Cohen's kappa which are described as follows:

Classification accuracy — The classification accuracy A_i and is calculated by the formula:

$$A_i = \frac{t}{n} \times 100 \qquad (16.4)$$

where t is the number of representatives correctly classified, and n is the total number of representatives.

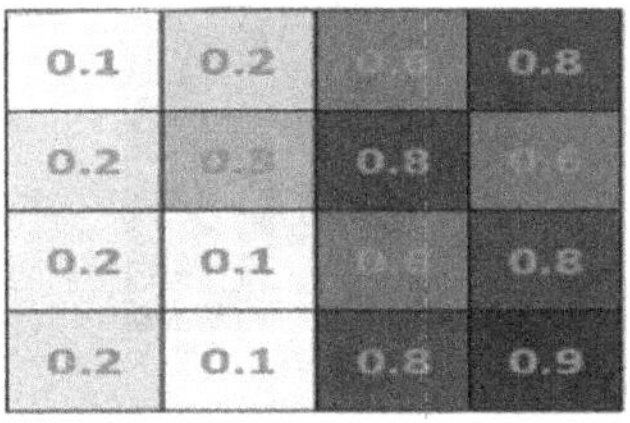

FIGURE 16.2 Real Values *Detected* on an Image.

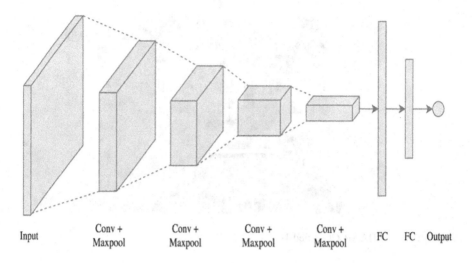

FIGURE 16.3 A Typical CNN Architecture.

Confusion matrix — A confusion matrix [53] is a table that is often used to express the performance of a classification model (or "classifier") on a set of test data for which the true values are known. Sensitivity and specificity are calculated as

$$Specificity = \frac{TN}{TN + FP}. \qquad (16.6)$$

F1 Score – F1 Score is a weighted average of the true positive rate (recall) and precision.

FIGURE 16.4 Image Classified as "Happy".

TABLE 16.3
Confusion Matrix for "Happy"

Total Number of Sample (N = 700)	Predicted Happy Yes	Predicted Not Happy No	Column Total
Actual happy Yes	140	35	175
Actual not happy No	30	495	525
Row total	170	530	

TABLE 16.4
Confusion Matrix for "Sad"

Total Number of Sample (N = 700)	Predicted Sad Yes	Predicted Not Sad No	Column Total
Actual sad Yes	114	61	175
Actual not sad No	60	465	525
Row total	174	526	

TABLE 16.5
Confusion Matrix for "Calm"

Total Number of Sample (N = 700)	Predicted Calm Yes	Predicted Not Calm No	Column Total
Actual calm Yes	132	43	175
Actual not calm No	40	485	525
Row total	172	528	

$$F_1 Score = 2 * \frac{\text{Precision} - \text{Recall}}{\text{Precision} + \text{Recall}} \quad (16.7)$$

Cohen's Kappa — It is a measure of how well the classifier performed compared to how well it would have performed simply by chance [54]. İt is calculated as Equation (16.5).

$$K = \frac{P_0 - P_e}{1 - P_e} \quad (16.8)$$

TABLE 16.6
Confusion Matrix for "Angry"

Total Number of Sample (N = 700)	Predicted Angry Yes	Predicted Not Angry No	Column Total
Actual angry Yes	125	50	175
Actual not angry No	47	478	525
Row total	172	528	

FIGURE 16.5 Image Classified as "Sad".

where P_0 is the relative observed agreement among raters, and P_e is the hypothetical probability of chance agreement, using the observed data to compute the probabilities of each observer randomly seeing each category.

The confusion matrices for different emotions are presented in Table 16.3 Tables 16.3 to 16.6 Table 16.6. The corresponding classified images are presented in Figures 16.5 to 16.7 Figure 16.5, Figure 16.6, Figure 16.7. The statistical information for validation of classification is shown in Table 16.7. Table 16.8 shows the statistical information for the validation of classification. From Table 16.8, it can be observed that the accuracies from different emotion classifications are quite good and acceptable. The overall accuracy lies in the range between 82-91%. Similarly, the misclassification rates lie in the range of 9-14%. The computed F1 scores are nominal and Cohen kappa measures are acceptable. These experimental results are in support of CNN-based classification.

The result obtained from our experimental study are described in Tables 16.7 and 16.8 using CNN. It is observed that overall accuracy of happy emotion is better than other emotions. The misclassification rate is less in happy emotion; it is higher in sad emotion.

FIGURE 16.6 Image Classified as "Calm".

FIGURE 16.7 Image Classified as "Angry".

6 CONCLUSION

Real-time facial emotion detection is a highly remarkable area of research in computer vision. CNN is the most accepted and successful deep learning architecture for image classification and reorganization when compared to conventional neural network. This chapter explored the state-of-the-art deep learning models for facial emotion recognition. We exploited CNN for facial emotion recognition in real-time with a video

TABLE 16.7
Statistical Information of Different Emotions from Confusion Matrix

Different Emotions	True Positive Rate (Recall): (TP/Actual Yes)	Precision (TP/Predicted Yes)	True Negative Rate (Specificity): TN/Actual No	False Positive Rate: FP/Actual No
Happy	0.8	0.82	0.9428	0.0571
Sad	0.6514	0.6551	0.8857	0.1142
Calm	0.7542	0.7674	0.9238	0.0761
Angry	0.7142	0.7267	0.9104	0.0895

TABLE 16.8
Performance Analysis of the Models through Different Measures

Different Emotions	Overall Accurcay (TP + TN)/Total = (Actual No + Actual Yes)	Misclassification Rate (FP + FN)/Total = (Actual No + Actual Yes)	F1 Score	Cohen's Kappa
Happy	90.71%	9.28%	0.0246	0.7499
Sad	82.71%	17.28%	0.0056	0.704
Calm	86.17%	11.85%	0.0022	1.2096
Angry	86.14%	13.85%	0.0173	0.8462

input stream. We developed a structure to detect human emotions in different scenes, angles, and lighting conditions real-time. Using our tradition trained network with a face-detector provided by OpenCV, we successfully execute the application of an emotion indicating one of four expressions (happy, sad, calm, and anger). We found that the performance metric values generated from the model are acceptable. Among four expressions, our model gives better result in "happy" expression. The present work can be extended by exploring other areas of computer vision.

REFERENCES

1. Krizhevsky, A., Sutskever, I., and Hinton, G. E. (2012). Imagenet classification with deep convolutional neural networks. In *Advances in Neural Information Processing Systems* (pp. 1097–1105).
2. Yao, A., Shao, J., Ma, N., and Chen, Y. (2015). Capturing Au-aware facial features and their latent relations for emotion recognition in the wild. In Proceedings of the 2015 ACM on International Conference on Multimodal Interaction (pp. 451–458).
3. Shan, C., Gong, S., and McOwan, P. W. (2009). Facial expression recognition based on local binary patterns: A comprehensive study. *Image and Vision Computing*, 27(6), 803–816.
4. Ebrahimi Kahou, S., Michalski, V., Konda, K., Memisevic, R., and Pal, C. (2015). Recurrent neural networks for emotion recognition in video. In Proceedings of the 2015 ACM on International Conference on Multimodal Interaction (pp. 467–474).

5. Yu, Z. and Zhang, C. (2015). Image based static facial expression recognition with multiple deep network learning. In Proceedings of the 2015 ACM on International Conference on Multimodal Interaction (pp. 435–442).
6. Kim, B. K., Roh, J., Dong, S. Y., and Lee, S. Y. (2016). Hierarchical committee of deep convolutional neural networks for robust facial expression recognition. *Journal on Multimodal User Interfaces*, *10*(2), 173–189.
7. Wu, Z., Peng, M., and Chen, T. (2016). Thermal face recognition using convolutional neural nnetwork. In 2016 International Conference on Optoelectronics and Image Processing (ICOIP) (pp. 6–9). IEEE.
8. Ramaiah, N. P., Ijjina, E. P., and Mohan, C. K. (2015). Illumination invariant face recognition using convolutional neural networks. In 2015 IEEE International Conference on Signal Processing, Informatics, Communication and Energy Systems (SPICES) (pp. 1–4). IEEE.
9. Lee, H. and Kwon, H. (2017). Going deeper with contextual CNN for hyperspectral image classification. *IEEE Transactions on Image Processing*, *26*(10), 4843–4855.
10. Hershey, S., Chaudhuri, S., Ellis, D. P., Gemmeke, J. F., Jansen, A., Moore, R. C., and Slaney, M. (2017). CNN architectures for large-scale audio classification. In 2017 IEEE International Conference on Acoustics, Speech and Signal Processing (ICASSP) (pp. 131–135). IEEE.
11. Wang, J., Yang, Y., Mao, J., Huang, Z., Huang, C., and Xu, W.,(2016). Cnn-rnn: A unified framework for multi-label image classification. In Proceedings of the IEEE Conference on Computer Vision and Pattern Recognition (pp. 2285–2294).
12. Han, D., Liu, Q., and Fan, W. (2018). A new image classification method using CNN transfer learning and web data augmentation. *Expert Systems with Applications*, *95*, 43–56.
13. Li, Q., Cai, W., Wang, X., Zhou, Y., Feng, D. D., and Chen, M.(2014). Medical image classification with convolutional neural network. In 2014 13th International Conference on Control Automation Robotics & Vision (ICARCV) (pp. 844–848). IEEE.
14. Spanhol, F. A., Oliveira L. S., Petitjean, C., and Heutte, L. (2015). A dataset for breast cancer histopathological image classification. *IEEE Transactions on Biomedical Engineering*, *63*(7), 1455–1462.
15. Wei, Y., Xia, W., Lin, M., Huang, J., Ni, B., Dong, J. and Yan, S. (2015). HCP: A flexible CNN framework for multi-label image classification. *IEEE Transactions on Pattern Analysis and Machine Intelligence*, *38*(9), 1901–1907.
16. Kang, L., Kumar, J., Ye, P., Li, Y., and Doermann, D. (2014). Convolutional neural networks for document image classification. In 2014 22nd International Conference on Pattern Recognition (pp. 3168–3172). IEEE.
17. Xie, S., Yang, T., Wang, X., and Lin, Y. (2015). Hyper-class augmented and regularized deep learning for fine-grained image classification. In Proceedings of the IEEE Conference on Computer Vision and Pattern Recognition (pp. 2645–2654).
18. Romero, A., Gatta, C., and Camps-Valls, G. (2015). Unsupervised deep feature extraction for remote sensing image classification. *IEEE Transactions on Geoscience and Remote Sensing*, *54*(3), 1349–1362.
19. Paoletti, M. E., Haut, J. M., Plaza, J., and Plaza, A.(2018). A new deep convolutional ceural network for fast hyperspectral image cassification. *ISPRS Journal of Photogrammetry and Remote Sensing*, *145*, 120–147.
20. Ioannou, Y., Robertson, D., Shotton, J., Cipolla, R., and Criminisi, (2015). A. training CNNS with low-rank filters for efficient image classification. arXiv preprint arXiv:1511.06744.
21. Hu, W., Huang, Y., Wei, L., Zhang, F., and Li, H. (2015). Deep convolutional neural networks for hyperspectral image classification. *Journal of Sensors*.
22. Affonso, C., Rossi, A. L. D., Vieira, F. H. A., and de Leon Ferreira, A. C. P. (2017). Deep learning for biological image classification. *Expert Systems with Applications*, *85*, 114–122.

23. Barat, C. and Ducottet, C. (2016). String representations and distances in deep convolutional neural networks for image classification. *Pattern Recognition, 54*, 104–115.
24. Li, B., Wang, K. C., Zhang, A., Yang, E., and Wang, G. (2020). Automatic classification of pavement crack using deep convolutional neural network. *International Journal of Pavement Engineering, 21*(4), 457–463.
25. Li, K., Jin, Y., Akram, M. W., Han, R., and Chen, J. (2020). Facial expression recognition with convolutional neural networks via a new face cropping and rotation strategy. *The Visual Computer, 36*(2), 391–404.
26. Arriaga, O., Valdenegro-Toro, M., and Plöger, P. (2017). Real-time convolutional neural networks for emotion and gender classification. arXiv preprint arXiv:1710.07557.
27. Matsugu, M., Mori, K., Mitari, Y., and Kaneda, Y. (2003). Subject independent facial expression recognition with robust face detection using a convolutional neural network. *Neural Networks, 16*(5-6), 555–559.
28. Jiang, H. and Learned-Miller, E. (2017, May). Face detection with the faster R-CNN. In 2017 12th IEEE International Conference on Automatic Face & Gesture Recognition (FG 2017) (pp. 650–657). IEEE.
29. Qin, H., Yan, J., Li, X., and Hu, X. (2016). Joint training of cascaded CNN for face detection. In Proceedings of the IEEE Conference on Computer Vision and Pattern Recognition (pp. 3456–3465).
30. Zhu, C., Zheng, Y., Luu, K., and Savvides, M. (2017). Cms-rcnn: Contextual multi-scale region-based CNN for unconstrained face detection. In *Deep Learning for Biometrics* (pp. 57–79). Springer, Cham.
31. Sun, X., Wu, P., and Hoi, S. C. (2018). Face detection using deep learning: An improved faster RCNN approach. *Neurocomputing, 299*, 42–50.
32. Hao, Z., Liu, Y., Qin, H., Yan, J., Li, X., and Hu, X. (2017). Scale-aware face detection. In Proceedings of the IEEE Conference on Computer Vision and Pattern Recognition (pp. 6186–6195).
33. Farfade, S. S., Saberian, M. J., and Li, L. J. (2015). Multi-view face detection using deep convolutional neural Networks. In Proceedings of the 5th ACM on International Conference on Multimedia Retrieval (pp. 643–650).
34. Li, H., Lin, Z., Shen, X., Brandt, J., and Hua, G. (2015). A convolutional neural network cascade for face detection. In Proceedings of the IEEE Conference on Comuputer Vision and Pattern Recognition (pp. 5325–5334).
35. Wang, H., Li, Z., Ji, X., and Wang, Y. (2017). Face R-CNN. arXiv preprint arXiv:1706.01061.
36. Zhang, K., Zhang, Z., Wang, H., Li, Z., Qiao, Y., and Liu, W. (2017). Detecting faces using inside cascaded contextual CNN. In Proceedings of the IEEE International Conference on Computer Vision (pp. 3171–3179).
37. Ranjan, R., Patel, V. M., and Chellappa, R. (2017). Hyperface: A deep multi-task learning framework for face detection, landmark localization, pose estimation, and gender recognition. *IEEE Transactions on Pattern Analysis and Machine Intelligence, 41*(1), 121–135.
38. Szlávik, Z., and Szirányi, T. (2004). Face analysis using CNN-UM. In Proceedings IEEE International Workshop on Cellular Neural Networks and Their Applications (CNNA 2004) (pp. 190–195).
39. Mori, K., Matsugu, M., and Suzuki, T. (2005, May). Face recognition using SVM fed with intermediate output of CNN for face detection. In *MVA* (pp. 410–413).
40. Raghavendra, R., Raja, K. B., Venkatesh, S., and Busch, C. (2017, July). Transferable deep-CNN features for detecting digital and print-scanned morphed face Images. In 2017 IEEE Conference on Computer Vision and Pattern Recognition Workshops (CVPRW) (pp. 1822–1830). IEEE.

41. Zhang, C., Xu, X., and Tu, D. (2018). Face detection using improved faster rcnn. arXiv preprint arXiv:1802.02142.
42. Yang, S., Luo, P., Loy, C. C., and Tang, X. (2015). From facial parts responses to face detection: A deep learning approach. In Proceedings of the IEEE International Conference on Computer Vision (pp. 3676–3684).
43. Wan, S., Chen, Z., Zhang, T., Zhang, B., and Wong, K. K. (2016). Bootstrapping face detection with hard negative examples. arXiv preprint arXiv:1608.02236.
44. Bansal, A., Castillo, C., Ranjan, R., and Chellappa, R. (2017). The do's and don'ts for CNN-based face verification. In Proceedings of the IEEE International Conference on Computer Vision Workshops (pp. 2545–2554).
45. Balya, D. and Roska, T. (1999). Face and eye detection by CNN algorithms. *Journal of VLSI Signal Processing Systems for Signal, Image and Video Technology*, *23*(2-3), 497–511.
46. Zhang, Z., Shen, W., Qiao, S., Wang, Y., Wang, B., and Yuille, A. (2020). Robust face detection via learning small faces on hard images. In The IEEE Winter Conference on Applications of Computer Vision (pp. 1361–1370).
47. Mori, K., Matsugu, M., and Suzuki, T. (2005). Face recognition using SVM fed with intermediate output of CNN for face detection. In *MVA* (pp. 410–413).
48. Khalajzadeh, H., Mansouri, M., and Teshnehlab, M. (2014). Face recognition using convolutional neural network and simple logistic classifier. In *Soft Computing in Industrial Applications* (pp. 197–207). Springer, Cham.
49. Ijjina, E. P. and Mohan, C. K. (2014). Facial expression recognition using kinect depth sensor and convolutional neural networks. In 2014 13th International Conference on Machine Learning and Applications (pp. 392–396). IEEE.
50. Connie, T., Al-Shabi, M., Cheah, W. P., and Goh, M. (2017). Facial expression recognition using a hybrid CNN–SIFT aggregator. In *International Workshop on Multi-disciplinary Trends in Artificial Intelligence* (pp. 139–149). Springer, Cham.
51. Luisier Wu, J. (2017). *Introduction to Convolutional Neural Networks. National Key Lab for Novel Software Technology*. Nanjing University. China, 5, 23.
52. Zha, S., Luisier, F., Andrews, W., Srivastava, N., and Salakhutdinov, R. (2015). Exploiting image-trained CNN architectures for unconstrained video classification. arXiv preprint arXiv:1503.04144.
53. https://www.dataschool.io/simple-guide-to-confusion-matrix-terminology/.
54. https://en.wikipedia.org/wiki/Cohen's_kappa.

Index

A

access control, 175, 178, 184–185
access key, 185
air pollution, 255
Amazon Web Service (AWS), 182, 184–186
American National Standards Institute (ANSI), 21
architecture, 19, 21, 25, 33, 90, 95–98, 226–227, 232
artificial neural network (ANN), 240
attack, 133–138, 140–145, 191–195, 200–202
Audi A3 E-Tron, 254
augmented reality (AR), 3, 5
Augmented Reality Mark-up Language (ARML), 5
asymmetric key cryptography, 171

B

base station, 31, 120, 123, 125, 156–157, 164, 192
battery EV, 253–254
battery-powered engine, 255
beacon, 192–193, 195–196
Bernoulli trials, 43
bidirectional, 22, 226, 229
big data systems, 32
bi-layer, 175–176, 187
binning, 52
binomial distribution, 43, 182
biodegradable materials, 31
Block With Holding (BWH), 174
blockchain, 97, 99, 102–104, 108, 165, 171–184, 187–189
BMW i8, 254
broadcast, 134, 138, 139
brown energy, 78
bucket, 185–187
business analytics, 38
business processing, 25

C

capability, 226
capacity of virtual machine (CVM), 63
carbon dioxide emission, 26, 28, 32, 60, 75
carbon footprint, 255
central, 223–224, 226, 229–231
channel identification, 118
Chevy Bolt, 254
Chevy Spark, 254
Chevy Volt, 254
Chrysler Pacifica, 254
classification, 47, 259, 265
cloud, 171–176, 178, 182–183, 187–189
cloud computing, 7, 20, 22, 26–28, 30, 33, 71–74, 77, 79–83, 79
cloud data center (CDC), 26–27, 75, 79, 84, 90
cloud environments, 25, 59, 73, 82–84, 90
cloud-resident capabilities, 72
cognitive, 100, 103, 105, 112–120, 123, 125–128, 130, 133–134, 143, 192, 205, 206, 220
Cognitive Base Station (CBS), 120–123
cognitive radio controller, 95–130
cognitive radio network (CRN), 113–130
Cognitive Relay Station (CRS), 120–123
Cohen's kappa, 267
collection, 48
common-use model, 114
communication, 226, 229, 231–232
Complementary Metal Oxide Semiconductor (CMOS), 30
confidence interval, 45
confusion matrix, 266, 270
consensus, 180–181
consumer, 225–226, 229
consumption, 231–232
continuous random variable, 43
Controllable Block Chain Data Management (CBDM), 173
conventional, 223–224, 230
convolutional neural network (CNN), 259–261, 264–265, 268–269
correlation, 47
critical value, 46–47
cryptography, 171, 174–175, 178, 188–189
CSDVN, 133–137, 139
CSV format, 48
current load (CL), 63

D

data, 133–136, 138, 143, 233
data analysis, 37, 100, 102
data exploration and transformation, 48
data extraction, 48

275

data plane, 133–135
data warehouse, 22, 227
database technology, 22, 25, 97, 112, 125, 127–128, 148, 158–159, 163–165, 174–176, 262–263
DDoS attacks, 9, 137, 161–162
decentralization, 226–231
decentralized applications (Dapps), 179
Delay Tolerant Networks (DTNs), 113–130
detection, 133–135, 138–140, 142–144
detection accuracy, 198, 200
detection of route, 85
detection time, 198–200
detrended fluctuation analysis (DFA), 240
digital signature, 103, 171
distributed, 226, 230, 233
distributed computing, 20, 25–26, 28, 59, 180
distributing system, 223
DoS, 9–10, 157, 160–162
dynamic spectrum access (DSA), 113–130

E

EEG, 259
electric vehicle, 253–257
electricity, 223–224, 226, 229–233
electrolytes, 255
electronic health record (EHR), 240
Electronic Product Environmental Assessment Tool (EPEAT), 21
electronic waste (e-waste), 19–20, 23
energy, 173, 180–182
energy bugs, 31
energy consumption, 198–200
energy consumption analysis, 4, 20, 22, 24–33, 60–62, 64, 72, 76, 82–90, 95, 97, 100, 139–142, 151, 153, 162, 181, 198–200, 229, 234
energy efficiency, 7, 20, 23, 27, 75, 83–84, 87, 89, 224, 230
Energy Efficient Request-Based Virtual Machine Placement, 76
energy grid, 109
entropy, 240
Ethereum, 178–179
Euclidean distance, 193
exclusive-use model, 114
execution time (ET), 63
exploitation, 180, 182

F

false negative, 264
false positive, 11, 264
feature extraction, 259

Federal Communication Commission (FCC), 112
fitness, 28, 63, 85, 87, 209, 213–214
F1 score, 266
Ford Focus Electric, 254
Ford Fusion Energi, 254
frequency table, 42

G

Ganache, 177, 179
gasoline engine, 255
genetic algorithm, 26–28, 60, 173, 206
genuine, 192, 194, 197, 199
global warming, 255
GPS, 191–197, 200
greedy, 241, 243
green design, 23
green disposal, 24
Green Energy Adaptive Interactive, 89
green figuring, 25
Green IT, 23
green manufacturing, 23
green mobile computing, 30
green use, 24
grid, 223–226, 228–235
grid map, 199
group, 134–135, 137–141

H

HAA, 260
HAAR, 261, 263
hardware technologies, 20
hash, 171, 179, 183–184
health monitoring, 95–130
heart disease prediction system (HDPS), 240
heart rate variability (HRV), 241
high-capacity batteries, 254
histogram, 37, 40, 52–53, 260
Hybrid Asymmetric Multicore Architecture (HAMA), 21
hybrid EV, 253–254
hydro, 223
hypothesis, 46–48, 263
hypothesis generation, 48

I

identification, 114
IEEE 802.11p, 139, 141
immutable, 180
incorrect optimistic rate (IOR), 12
Indian, 223–235
inferential statistics, 37, 44

Index

Infrastructure-as-a-Service (IaaS), 71
integrate, 22, 223, 226, 231
intelligent, 229–231
interface, 135–136, 140, 194, 198
internal combustion engines, 253–255
Internet of Things (IoT), 1, 4–5, 7–11, 14, 20, 31, 33, 97–98, 102–105, 111, 117, 135, 141, 147–165, 182, 193–200, 223–235
interstitial lung disease (ILD), 260
interquartile range, 40
intrusion detection system (IDS), 12
IoT + AR, 5, 9
IT services, 26, 72

J

job scheduling, 26, 28, 61, 75

K

K-Nearest Neighbour (KNN), 241
K-mean clustering, 239, 241–244, 249

L

lambda, 183
latency, 186
layer, 171, 175–178, 187
LC, 134, 136, 138
leader, 134, 137
light motor vehicles, 256
Lithium-ion batteries, 255–256
load, 225–231
load balancing, 27–28, 59–67, 75, 79, 90, 174
load Information at the virtual machine (LVM), 63
Local Binary Patterns (LBP), 260
long term evolutionary (LTE), 30
lubrication, 255

M

machine learning (ML), 3, 47, 53, 55, 76, 100, 103, 117, 134, 165, 192, 224, 226, 235, 239–241, 243, 245
machine-to-machine communication (M2M), 111
makespan of the virtual machine (MVM), 63
makespan time calculation, 66
malicious, 139–140, 144, 192, 196–197, 199–200
Management Information System (MIS), 38
Map Reading Routing (MRR), 86–90
margin of error, 45
mean, 39

median, 39
meter, 231–233
microgrid, 228–229, 232, 234
miniaturization board level, 23
Minimizing Overhead Problem (MOP), 121
misbehavior, 133–137, 140–141
mobile cloud computing, 31
Mobile Internet of Things (MIoT), 4
mobility, 114
mode, 39
model construction, 76
model practice, 76
monitor, 223–224, 226, 228, 230–231
multilayer perceptron (MLP), 260, 264

N

national, 229, 232–233
neighbor, 133–135, 137–140, 191–193, 195–197
network, 133–135, 137–138, 140–145, 174, 176, 179, 184, 187
Next Generation Network, 111–113
Nissan LEAF, 254
node, 133, 135, 140–141, 143–144
nonce, 180–181
Non-Real Time User (NRT), 116
normal distribution, 43

O

OBU, 135
OMNeT++, 134, 140, 145, 192, 198–199, 202
optimal, 224
optimization, 171, 173, 178, 186, 188
Optimization Model, 76
Optimized Channel Selection Scheme, 95–130
outlay optimization, 88
outliers, 37–41, 49, 51–52

P

packet, 135–137, 141, 193–195, 198
peer-to-peer, 176, 189
pillar, 225
public keys, 171
platform-as-a-service (PaaS), 71
platooning, 133–137, 140–143
plug-in hybrid EV, 253–255
position, 133, 135, 137, 144, 191–193, 195–196, 202
post disaster management, 95–130
potential, 223, 226, 229
power, 223–226, 228–235
Power Optimization Context, 6

predictive modelling, 37–38, 48, 53
Principal Component Analysis (PCA), 241
private key, 175, 178
private key generator (PKG), 173
Probability Density Function, 45
problem definition, 48
processing time, 186
Proof of Work (PoW), 180–182
public key, 171, 175, 178
puzzle, 180–181
pyramid histogram of gradient (PHOG), 260
Python libraries and functions, 55
Python programming, 39

Q

QoS algorithm, 112

R

radial basis function (RBF), 244–245
random variable, 42
Real-Time User (RT), 116
recital guarantee via edition, 88
regenerative braking, 254
regression, 47
reliable, 224–232
reliable link quality, 121
remote, 231–232
renewable, 223–224, 229–230, 233–234
renewable energy, 26, 71–73, 75, 77–81, 83, 90, 95–97, 106, 109, 223–225, 230
result, 192, 197, 199
RFID, 13, 31, 149–150, 153
Round Trip Time (RTT), 88–89
routing, 133, 135, 143–145
routing process, 120–130
RSSI, 192, 196
RSU, 133–141, 191–193, 196–198

S

safety, 133, 135, 137–138
scenario, 223–224
SDN controller, 133–134, 136, 138, 141
SDVN, 192–193, 200
secret key, 173, 185
sensing, 114
sensor, 224, 226–227, 230–233
service, 171–172, 176, 178–179
SGM, 138–139
shared-use model, 114
sidechain, 177

similarity, 138–140
simulation, 65, 192, 197–200, 202
small-scale routing algorithm, 84–90
smart, 223–235
smart contracts, 177
smart grid, 224–226, 229–231, 234
smart portable devices, 30
smart transportation, 13
smartphone applications, 30
software, 228, 231–233
software-as-a-service (SaaS), 71
software technologies, 20
solar energy, 255
solar PV model, 80
solidity, 177
solid-state batteries, 256
sports utility vehicles, 256
spread of data, 41
standard deviation, 42
statistical methods, 37–57, 205
S3, 183–185
storage, 171–173, 175, 183–184
substation, 223, 226, 230
SUMO, 140–141
Super Video Graphics Array (SVGA), 21
supervised learning, 47
support vector machine (SVM), 239, 241–244, 246, 249
sustainable green computing, 70
Sustainable Green Energy Cloud Data Centre, 79

T

table, 193, 195, 197–199
task scheduling, 26–27, 59, 61, 64, 75, 112
TB-PAD, 133–145
technology, 223–235
Tesla X, 254
THD, 223–224
tidal energy, 256
transaction, 171, 175, 177, 179–180
transmission, 223–235
true positive, 247
Truffle, 179
trust, 133–140, 143–144
t-tests, 46
two-layer, 178, 186–187

U

unsupervised learning, 47
user, 173–175, 178–180, 182–184
utilities, 22, 224, 226, 229
utilization, 226, 229

Index

V

validation, 76
VANET, 133–135, 141, 143–144, 191–193, 200–202
variable transformation, 52
variables, 42
variance, 41
vehicle, 133–141, 144, 191–202
Video Electronics Standard Association (VESA), 21
virtual reality (VR), 6, 14
Virtual Reality Geographic Information System (VRGIS), 6
Virtual Reality Technology and Terrestrial Statistics Structure, 6
virtualization, 25–27, 59–60, 71, 76, 82–84, 174
Volkswagen e-Golf, 254

V2I, 133, 135, 191, 193
V2V, 133, 135, 191, 193

W

wind energy, 256
Wireless Body Area Network (WBAN), 112–115
wireless sensor networks (WSN), 31–32, 87, 124, 127, 135, 141, 148, 150, 157–158, 161, 193, 200

Z

Z-score, 44, 46

Printed in the United States
by Baker & Taylor Publisher Services